Pollution Control in Oil, Gas and Chemical Plants

Alireza Bahadori

Pollution Control in Oil, Gas and Chemical Plants

Springer

Alireza Bahadori
School of Environment, Science and Engineering
Southern Cross University
East Lismore, NSW
Australia

ISBN 978-3-319-01233-9 ISBN 978-3-319-01234-6 (eBook)
DOI 10.1007/978-3-319-01234-6
Springer Cham Heidelberg New York Dordrecht London

Library of Congress Control Number: 2013946885

Printed on acid-free paper

Springer is part of Springer Science+Business Media (www.springer.com)

Preface

The increasing energy demand over the past decades has resulted in a corresponding growth and expansion in the refining and processing of crude petroleum, natural gas, and many other chemical industries. The proliferation of refineries and chemical plants, combined with the increasingly stringent quality requirements for effluents discharged from these facilities, has underscored the need for perfecting existing pollution control technology and developing improved approaches for minimizing the pollution potential in this sector.

Exploration and production of natural gas and oil are rapidly expanding across the world due to technological developments that have made extraction of previously untapped unconventional resources, such as shale gas, economically feasible. In part due to expansion of operations to locations where the industry has never been active, emissions of air pollutants, along with other potential environmental and public health impacts, have come under increasing scrutiny.

Industries and municipalities account for the largest portion of waste discharges that occur. Industries discharge an enormous variety of materials which include the largest volume of waste and the most toxic of pollutants. Environmental control is a very complex science. The trend is to move in the direction of pollution avoidance rather than control.

This will add a tremendous cost to industrial budget. The public has tolerated pollution until recently, but the condition has increased to the point that many may suffer or even die because of it especially in developing countries.

In this book monitoring of wastewater control, equipment, and engineering techniques will be discussed. This book specification is intended to cover the safety and environmental control aspects as the essential requirements for water pollution control in oil, gas and chemical plants. The scope is accomplished under following titles:

- Refinery, gas processing industries, petrochemical plants, and oil terminal water pollution control.
- Chemical industries, water pollution sources and control.
- Organic chemical manufacturing.
- Monitoring.

Cleaning up hydrocarbon-based liquids and refined products release from pipelines, oil production units, pumping station, tank farms, and drilling operation in upstream part and other pollutants from oil and petrochemical refineries and above ground/underground storage tanks in downstream part of both petroleum and chemical industries, typically involves using several corrective actions and strategies. Short-term emergency measures may involve imminent actions to control acute safety and health hazards such as potential explosion and toxication. After the imminent danger has been eliminated, longer term corrective actions involve cleaning up pollutants that have entered the surface and subsurface environment.

In this book the following matter of *Soil Pollution Control* is discussed:

- To provide authorities concerned on how to assess site conditions in the unsaturated zone and where a petroleum product release has occurred, information needed to localize where in the unsaturated zone petroleum product is located, and also the removal of petroleum products from the unsaturated zone at a given site.
- To assess on the technologies designed specifically for clean-up of the saturated zone.
- To provide a structural methodology for evaluation and potential consequences of a leak in a pipeline. The methodology that is intended to assist pipeline operators in assessing the need to install pipeline leak detection facilities, and an overview of available pipeline leak detection techniques.

Petroleum products in the subsurface may be trapped between soil particles in the unsaturated zone, floating on the water table or dissolved in ground water in the saturated zone. The focus of this book is on long-term strategies for cleaning up petroleum products in unsaturated or saturated zones. When the spill occurs some kind of soil clean-up is typically necessary.

Air pollution is waste remaining from the many ways as production of goods, process of hydrocarbon and petrochemicals, use of transport, and generation of energy to heat and light. The major cause of all air pollution is combustion. When perfect or theoretical combustion occurs, the hydrogen and carbon in the fuel combine with oxygen from the air to produce heat, light, carbon dioxide, and water vapor. Impurities in the fuel, poor fuel-to-air ratio, or too high or too low combustion temperatures cause the formation of such side products as carbon monoxide, sulfur oxides, nitrogen oxides, fly ash, and unburned hydrocarbons. In this book, methods of reducing air pollution and means to decrease the pollution of the environment is discussed.

This book represents the essential requirements for air pollution control and covers the following main topics:

- Pollution sources
- Types of emission
- Measurement of air pollution
- Consideration in the selection of the pollution control equipment

- Process control
- Threshold limit values (TLV).

The effects of air pollution on human health, on plants and animals are well known. To ensure that all important methods of emission reduction have been considered, this book is prepared to provide information which will be of values to all concerned with measurements of general pollution of the air and involved in reducing to a realistic minimum the emission from industrial processes. It is an advantage that there are accepted methods for the measurement so that a sound comparison of all results could be made.

Due to the various adverse impacts of noise on humans and environment, noise should be controlled. The technique or the combination of techniques to be employed for noise control depend upon the extent of the noise reduction required, nature of the equipment used, and the economy aspects of the available techniques.

This book provides new insights and information with regard to reduction in the noise exposure time or isolation of species from the noise sources. It covers noise pollution control techniques besides providing personal ear protection, engineered control for noise reduction at source, and/or diversion in the trajectory of sound waves.

Last but not least I would like to sincerely thank the Springer Editorial Team, Michael Luby and Merry Stuber, for their editorial assistance, advices and support during working on this book project.

Contents

Air Pollution Control

1

Abstract

Petroleum and chemical processes are responsible for many emissions both into the air. Most relevant emissions into the air are nitrous oxides (NOx), sulfur oxides (SOx), carbon monoxide (CO), methane, and volatile organic compounds (VOC).The major cause of all air pollution is combustion. When perfect or theoretical combustion occurs, hydrogen and carbon in the fuel combine with oxygen from the air to produce heat, light, carbon dioxide, and water vapor. Impurities in the fuel, poor fuel-to-air ratio, or too high or too low combustion temperatures cause the formation of such side products as carbon monoxide, sulfur oxides, nitrogen oxides, fly ash, and unburned hydrocarbons. In this chapter, methods of reducing air pollution and means to decrease the pollution of the environment are discussed.

Keywords

Air • Clean air. Emission sources • Flare • Fugitive emissions • Odor control • Organic emissions oxides of nitrogen (NOx) • Particulate matter • Pollution • Threshold limit values • Volatile organic compounds (VOCs)

Air pollution is the contamination of the air by noxious gases and minute particles of solid and liquid matter (particulates) in concentrations that endanger health. In this condition, air is contaminated by foreign substances, or the substances themselves.

Serious damage can be caused by polluting material discharged into environment from various sources. Such emission can adversely affect agricultural land and crops and can spoil clothing and buildings, as well as corrode metal installations and can have harmful effect on the health of community.

Air pollution is waste remaining from the many ways as production of goods, process of hydrocarbon and petrochemicals, use of transport and generation of energy to heat and light.

A. Bahadori, *Pollution Control in Oil, Gas and Chemical Plants*,

DOI: 10.1007/978-3-319-01234-6_1,

Air pollution consists of gaseous, liquid, or solid substances that, when present in sufficient concentration, for a sufficient time, and under certain conditions, tend to interfere with human comfort, health, or welfare and cause environmental damage. Air pollution causes acid rain, ozone depletion, photochemical smog, and other such phenomena.

This chapter represents the minimum requirements for air pollution control and covers the following main topics:

- Pollution sources
- Types of emission
- Standard levels
- Measurement of air pollution
- Consideration in the selection of the pollution control equipment
- Process control
- Threshold limit values (TLV)

1.1 Air Pollution Definition

Equipment leaks sometimes in chemical and petroleum processing industries are responsible for significant amount of emissions. Even if each individual leak is generally small, it is the largest source of emissions of volatile organic compounds (VOCs) and volatile hazardous air pollutants (VHAPs) from petroleum industries (in particular refineries) and chemical manufacturing facilities.

Air pollution is defined as the addition to our atmosphere of any material which will have a deleterious effect to life upon our planet. The material might be a toxic gaseous hydrocarbon with some long-lasting effect on an organism ingesting it or perhaps a particulate irritant which could cause similar problems. It might be atomic radiation in a form that we cannot see which would be damaging to animal or plant cells. A pollutant can be anything which when put into the atmosphere either purposely or through some act of nature reduces the oxygen content or significantly changes the composition of the air.

An air pollutant does not have to be inhaled. It becomes a pollutant merely by being in the air. Smog, made up of gases and particulate matter in large quantities, and seen over many of our major cities in the last 15 years, forms a blanket which shields some of the sun's radiation so necessary to life on earth. This blanket changes heat absorption patterns on the surface of the earth and can prohibit reradiation of the earth's heat to the sky, resulting in the "greenhouse effect" which ultimately changes weather and temperature patterns across the entire surface of the planet.

1.2 Clean Air

The earth is surrounded by approximately 500 billion tons of air, of which 20.9 % is oxygen, 78.0 % is nitrogen, and 0.9 % is argon. Excluding water vapor, these gases make up all but 0.04 % of the atmosphere and three-quarters of the remainder is carbon dioxide. The last 0.01 % contains the other noble gases, hydrogen, nitrogen oxides, ozone, and traces of many other gases. Table 1.1 summarizes the major gases and their concentrations on a dry basis.

These figures remain essentially constant as long as water vapor is not included, since it varies in a range from as high as 3 % by volume to less than 1 %. In ambient air measurements, this variation is normally within the reproducibility limits of the determinations and can be ignored. Despite this, samples are often dried before testing.

Water vapor normally cannot be considered as an air pollutant even though it is produced in great quantities in combustion processes. Natural evaporation and precipitation move far greater amounts of water than any man-made process. One exception to this may occur in the near future with the advent of the supersonic transport. This plane may release large amounts of water vapor into the stratosphere causing increased cloudiness and a temperature rise.

Table 1.1 Approximate concentration (volume %)

Constituent	Approximate concentration
Nitrogen (N_2)	78.03
Oxygen (O_2)	20.99
Carbon dioxide (CO_2)	0.03
Argon (A)	0.94
Neon (Ne)	0.00123
Helium (He)	0.0004
Krypton (Kr)	0.00005
Xenon (Xe)	0.000006
Hydrogen (H_2)	0.01
Methane (CH_4)	0.0002
Nitrous oxide (N_2O)	0.0005
Water vapor (H_2O)	Variable
Particulate matter	Variable type and quantity (the types and concentrations of these substances may vary substantially from one region to another and within any period of time from natural conditions)
Ozone (O_3)	Variable (from ultraviolet radiation and probably thunderstorms, concentrations will vary from 0 to 0.07 ppm)
Formaldehyde (HCHO)	Uncertain (from biological sources or oxidation of CH_4, possible concentrations are not determined)

Water vapor has a synergistic effect upon air pollutants, increasing many deleterious effects. For example, sulfur dioxide, which itself is an irritating and corrosive pollutant, can combine with water vapor and atmospheric oxygen to form highly corrosive sulfuric acid mist. Smog also depends upon water vapor for its formation. Monitoring of pollution in urban areas must consider the humidity and predict its effect on pollutants.

1.3 Pollution Sources

- Major categories of air pollution are considered to be caused by:
  1. Transportation
  2. Domestic heating
  3. Electric power generation
  4. Industrial fuel-burning and process emission.
- Major industrial polluters have been classified relative to the type of industry. Generally, they can be categorized as follows:
  (a) Petroleum refineries; particulate, sulfur oxide, hydrocarbons, and carbon monoxide.
  (b) Petrochemical industries; hydrocarbon, carbon monoxide particulate.
  (c) Fertilizer industries particulates, NOx, gaseous fluoride compounds.

1.4 Types of Emissions in Petroleum Industries

1.4.1 Refinery

Potential sources of refinery emissions are given in Table 1.2.

1.4.2 Atmospheric Pollution in Refineries

The average emission factor approach is a combination of average emission factors and unit-specific data: number of each type of equipment (valves, pump seals, etc.), the service each equipment is in (gas, light liquid, heavy liquid), the total organic compound (TOC) concentration of the stream, and time period each equipment is in that service. The emission rate of TOC from all equipment can be calculated with the following formula:

$$E_{TOC} = F_A \times WF_{TOC} \times N \tag{1.1}$$

where

E_{TOC} emission rate of TOC from all equipments in the stream of a given equipment type (kg/h)

Table 1.2 Potential sources of specific emissions from oil refineries

Types of emission	Potential sources
Oxides of sulfur	Boilers, process heaters, catalytic cracking unit regenerators, treating units, H_2S flares, decoking operations, etc
Hydrocarbons	Loading facilities, turnarounds, sampling, storage tanks, wastewater separators, blowdown systems, catalyst regenerators, pump, valves, blind changing, cooling towers, vacuum jets, barometric condensers, air blowing, high-pressure equipment handling volatile hydrocarbons, process heaters, boilers, compressor engines, etc
Oxides of nitrogen	Process heaters, boilers, compressor engines, catalyst regenerators, flares, etc
Particulate matter	Catalyst regenerators, boilers, process heater decoking operations, incinerators, etc
Aldehydes	Catalyst regenerators
Ammonia	Catalyst regenerators
Odors	Treating units (air blowing, steam blowing) drains, tank vents, barometric condenser sumps, wastewater separators
Carbon monoxide	Catalyst regeneration, decoking, compressor engines, incinerators
Arsine	Regeneration hydrodesulfurized
Phosgene	Regeneration hydrodesulfurized
Cyanides	Regeneration hydrodesulfurized
HCl	Platformers are omerization
HF	HF alkylation
H_2SO_4	Alkylation acid plant

F_A applicable average emission factor for the equipment type (kg/h/source)
WF_{TOC} average weight fraction of TOC in the stream
N number of pieces of equipment of the applicable equipment type in the stream

Tables 1.3, 1.4, 1.5 gather all the average emission factors. Synthetic organic chemical manufacturing industry (SOCMI) factors are used to determine equipment leak emissions from chemical plants or chemical processes within refineries. These factors are for TOC emission rates.

Emission factor for major atmospheric pollution in refinery is given in Table 1.4.

Refinery factors are used to determine equipment leak fugitive emissions from a refinery process. For a chemical process located within a refinery that is not specifically considered a refinery process (for example, an MTBE production unit),

Table 1.3 Emission factor for synthetic organic chemical manufacturing industry (SOCMI)

Equipment type	Service	Emission factor (kg/h/source)
Valves	Gas	0.00597
	Light liquid	0.00403
	Heavy liquid	0.00023
Pump seals/Agitator seals	Light liquid	0.0199
	Heavy liquid	0.00862
Compressor seals	Gas	0.228
Pressure relief valves	Gas	0.104
Connectors	All	0.00183
Open-ended lines	All	0.0017
Sampling connections	All	0.0150

the SOCMI factors must be used rather than the refinery factors. These factors are for non-methane organic compound emission rates.

These factors in Table 1.5 are for TOC emission rates.

- Example calculation:

Assuming a process operating gaseous product, calculation of fugitive emissions can be done like following. Data for example calculation are given in Table 1.6.

The below formulas are used in the calculations to obtain the results in Table 1.7:

Table 1.4 Refinery average emission factors

Equipment type	Service	Emission factor (kg/h/source)
Valves	Gas	0.0268
	Light liquid	0.0109
	Heavy liquid	0.0023
Pump seals/Agitator seals	Light liquid	0.114
	Heavy liquid	0.021
Compressor seals	Gas	0636
Pressure relief valves	Gas	0.16
Connectors	All	0.00025
Open-ended lines	All	0.0023
Sampling connections	All	0.015

Table 1.5 Oil and gas production operations average emission factors

Equipment type	Service	Emission factor (kg/h/source)
Valves	Gas	4.5 E-03
	Heavy oil	8.4 E-06
	Light oil	2.5 E-03
	Water/Oil	9.8 E-05
Pump seals	Gas	2.4 E-03
	Heavy oil	Not available
	Light oil	1.3 E-02
	Water/Oil	2.4 E-05
Others	Gas	8.8 E-03
	Heavy oil	3.2 E-05
	Light oil	7.5 E-03
	Water/Oil	1.4 E-02
Connectors	Gas	2.0 E-04
	Heavy oil	7.5 E-06
	Light oil	2.1 E-04
	Water/Oil	1.1 E-04
Flanges	Gas	3.9 E-04
	Heavy oil	3.9 E-07
	Light oil	1.1 E-04
	Water/Oil	2.9 E-06
Open-ended lines	Gas	2.0 E-03
	Heavy oil	1.4 E-04
	Light oil	1.4 E-03
	Water/Oil	2.5 E-04

- VOC emissions rate = (equipment count) (wt. fraction) (emission factor)
- Total VOC emissions = sum [(VOC emissions rate) (hours of operation)]

Total TOC fugitive emission from a unit process can be known by summing emissions from each type of components, from each stream. Average factors generally determine total hydrocarbon emissions. In order to determine total VOC emissions, the calculated emission rates must be multiplied by the stream's weight percentage of VOC compounds (Indeed, it can happen that not all organic compounds present in the stream be classified as VOCs, for instance methane or ethane).

Table 1.6 Given data for this example

Equipment type/service	Equipment count	Hours of operation (h/yr) (hours of operation include all time in which material is contained in the equipment)	VOC wt. fraction
Valves/gas	272	0.0268	7.2896
Compressors/gas	3	0.636	1.908
Relief valves	37	0.16	5.92
Open-ended lines	489	0.0023	1.1247
Sampling connections	24	0.015	0.36

Table 1.7 Calculated results

Source	Equipment count	Emission factor (kg/h/source)	VOC emissions rate (kg/h)
Valves	272	0.0268	7.2896
Compressor seals	3	0.636	1.908
Relief valves	37	0.16	5.92
Open-ended lines	489	0.0023	1.1247
Sampling connections	24	0.015	0.36
Total VOC emissions (t/yr)			145.4

If some of the organic compounds in the stream are not classified as VOCs, total VOCs emission can be calculated with the following formula:

$$E_{VOC} = E_{TOC} \times (WP_{VOC}/WP_{TOC}) \tag{1.2}$$

where

E_{VOC} the VOC mass emissions from the equipment (kg/h)
E_{TOC} the TOC mass emissions from the equipment (kg/h)
WP_{VOC} the VOC concentration in the equipment in weight percent
WP_{TOC} the TOC concentration in the equipment in weight percent.

If, for a stream, estimating emissions of a specific VOC in the mixture is necessary, the following formula can be used:

$$E_{X} = E_{TOC} \times (WP_{X}/WP_{TOC}) \tag{1.3}$$

where

E_X	the organic chemical "X" mass emissions from the equipment (kg/h)
E_{TOC}	the TOC mass emissions from the equipment (kg/h)
WP_X	the organic chemical "X" concentration in the equipment in weight percent
WP_{TOC}	the TOC concentration in the equipment in weight percent.

Three other methods emitted from the *protocol for equipment leak emission estimates* are available. However, these methods necessitate on-site monitoring so they are not included in the scope of this study.

The following process emission sources are discussed here:

1. Vacuum distillation
2. Catalytic cracking
3. Thermal cracking processes
4. Utility boilers
5. Heaters
6. Compressor engines
7. Blowdown systems
8. Sulfur recovery.

1.4.3 Vacuum Distillation Emission Sources

Topped crude withdrawn from the bottom of the atmospheric distillation column is composed of high boiling point hydrocarbons. When distilled at atmospheric pressures, the crude oil decomposes and polymerizes and will foul equipment. To separate topped crude into components, it must be distilled in a vacuum column at a very low pressure and in a steam atmosphere.

The major sources of atmospheric emissions from the vacuum distillation column are associated with the steam ejectors or vacuum pumps. A major portion of the vapors withdrawn from the column by the ejectors or pumps is recovered in condensers. Historically, the non-condensable portion of the vapors has been vented to the atmosphere from the condensers. There are approximately 0.14 kg of non-condensable hydrocarbons per m^3 of topped crude processed in the vacuum distillation column.

A second source of atmospheric emissions from vacuum distillation columns is combustion products from the process heater. Fugitive hydrocarbon emissions from leaking seals and fittings are also associated with the vacuum distillation unit, but these are minimized by the low operating pressures and low vapor pressures in the unit.

Control technology applicable to the non-condensable emissions vented from the vacuum ejectors or pumps includes venting into blowdown systems or fuel gas systems, and incineration in furnaces or waste heat boilers. These control techniques are generally greater than 99 % efficient in the control of hydrocarbon emissions, but they also contribute to the emission of combustion products.

1.4.4 Catalytic Cracking Emission Sources

Catalytic cracking, using heat, pressure, and catalysts, converts heavy oils into lighter products with product distributions favoring the more valuable gasoline and distillate blending components. Feedstocks are usually gas oils from atmospheric distillation, vacuum distillation, coking, and deasphalting processes. These feedstocks typically have a boiling range of 340–540 °C. All of the catalytic cracking processes in use today can be classified as either fluidized bed or moving bed units.

Air emissions from catalytic cracking processes are (1) combustion products from process heaters and (2) flue gas from catalyst regeneration. Emissions from process heaters are discussed below. Emissions from the catalyst regenerator include hydrocarbons, oxides of sulfur, ammonia, aldehydes, oxides of nitrogen, cyanides, carbon monoxide (CO), and particulates. The particulate emissions from FCC units are much greater than those from TCC units because of the higher catalyst circulation rates used.

FCC particulate emissions are controlled by cyclones and/or electrostatic precipitators. Particulate control efficiencies are as high as 80–85 %. Carbon monoxide waste heat boilers reduce the CO and hydrocarbon emissions from FCC units to negligible levels. TCC catalyst regeneration produces similar pollutants to FCC units, but in much smaller quantities.

The particulate emissions from a TCC unit are normally controlled by high-efficiency cyclones. Carbon monoxide and hydrocarbon emissions from a TCC unit are incinerated to negligible levels by passing the flue gases through a process heater firebox or smoke plume burner. In some installations, sulfur oxides are removed by passing the regenerator flue gases through a water or caustic scrubber.

1.4.5 Thermal Cracking Emission Sources

Thermal cracking processes include visbreaking and coking, which break heavy oil molecules by exposing them to high temperatures.

From available literature, it is unclear what emissions are released and where they are released. Air emissions from thermal cracking processes include coke dust from decoking operations, combustion gases from the visbreaking and coking process heaters, and fugitive emissions. Emissions from the process heaters are discussed below.

Fugitive emissions from miscellaneous leaks are significant because of the high temperatures involved and are dependent on equipment type and configuration, operating conditions, and general maintenance practices.

Particulate emissions from delayed coking operations are potentially very significant. These emissions are associated with removing the coke from the coke drum and subsequent handling and storage operations. Hydrocarbon emissions are also associated with cooling and venting the coke drum before coke removal. However, comprehensive data for delayed coking emissions have not been included in available literature.

Particulate emission control is accomplished in the decoking operation by wetting down the coke. Generally, there is no control of hydrocarbon emissions from delayed coking. However, some facilities are now collecting coke drum emissions in an enclosed system and routing them to a refinery flare.

1.4.6 Blowdown System Emission Sources

The blowdown system provides for the safe disposal of hydrocarbons (vapor and liquid) discharged from pressure relief devices. Uncontrolled blowdown emissions primarily consist of hydrocarbons but can also include any of the other criteria pollutants. The emission rate in a blowdown system is a function of the amount of equipment manifolded into the system, the frequency of equipment discharges, and the blowdown system controls.

Emissions from the blowdown system can be effectively controlled by combustion of the non-condensables in a flare. To obtain complete combustion or smokeless burning (as required by most states), steam is injected in the combustion zone of the flare to provide turbulence and air. Steam injection also reduces emissions of nitrogen oxides by lowering the flame temperature.

1.4.7 Process Heaters Emission Sources

Process heaters (furnaces) are used extensively in refineries to supply the heat necessary to raise the temperature of feed materials to reaction or distillation level.

All the criteria pollutants are emitted from process heaters. The quantity of these emissions is a function of the type of fuel burned, the nature of the contaminants in the fuel, and the heat duty of the furnace. Sulfur oxides can be controlled by fuel desulfurization or flue gas treatment. Carbon monoxide and hydrocarbons can be controlled by more combustion efficiency. Currently, four general techniques or modifications for the control of nitrogen oxides are being investigated: combustion modification, fuel modification, furnace design, and flue gas treatment. Several of these techniques are being applied to large utility boilers, but their applicability to process heaters has not been established.

1.4.8 Compressor Engines Emission Sources

Many older refineries run high-pressure compressors with reciprocating and gas turbine engines fired with natural gas. The major source of emissions from compressor engines is combustion products in the exhaust gas. These emissions include CO, hydrocarbons, nitrogen oxides, aldehydes, and ammonia.

Sulfur oxides may also be present, depending on the sulfur content of the natural gas. All these emissions are significantly higher in exhaust from reciprocating engines than from turbine engines. The major emission control technique applied to compressor engines is carburetion adjustment similar to that applied on automobiles. Catalyst systems similar to those of automobiles may also be effective in reducing emissions, but their use has not been reported.

1.4.9 Sweetening Emission Sources

Sweetening of distillates is accomplished by the conversion of mercaptans to alkyl disulfides in the presence of a catalyst. The major emission problem is hydrocarbons from contact of the distillate product and air in the "air blowing" step. These emissions are related to equipment type and configuration, as well as to operating conditions and maintenance practices.

1.4.10 Asphalt Blowing Emission Sources

The asphalt blowing process polymerizes asphaltic residual oils by oxidation, increasing their melting temperature and hardness to achieve an increased resistance to weathering.

Air emissions from asphalt blowing are primarily hydrocarbon vapors vented with the blowing air. The quantities of emissions are small because of the prior removal of volatile hydrocarbons in the distillation units, but the emissions may contain hazardous polynuclear organics. Emissions are 30 kg/megagram (Mg) of asphalt. Emissions from asphalt blowing can be controlled to negligible levels by vapor scrubbing, incineration, or both.

1.4.11 Fugitive Emissions and Controls

Fugitive emission sources include leaks of hydrocarbon vapors from process equipment and evaporation of hydrocarbons from open areas, rather than through a stack or vent. Fugitive emission sources include valves of all types, flanges, pump and compressor seals, process drains, cooling towers, and oil/water separators. Fugitive emissions are attributable to the evaporation of leaked or spilled petroleum liquids and gases. Normally, control of fugitive emissions involves minimizing leaks and spills through equipment changes, procedure changes, and improved monitoring, housekeeping, and maintenance practices.

1.4.12 Valves, Flanges, Seals, and Drains Emission Sources

For these sources, a very high correlation has been found between mass emission rates and the type of stream service in which the sources are employed. Valves, because of their number and relatively high emission factor, are the major emission source. This conclusion is based on an analysis of a hypothetical refinery coupled with the emission rates. The total quantity of fugitive VOC emissions in a typical oil refinery with a capacity of 52,500 m^3 per day is estimated as 20,500 kg per day.

1.4.13 Wastewater Treatment Plant Emission Sources

All refineries employ some form of wastewater treatment so water effluents can safely be returned to the environment or reused in the refinery. The main components of atmospheric emissions from wastewater treatment plants are fugitive VOCs and dissolved gases that evaporate from the surfaces of wastewater residing in open process drains, separators, and ponds.

1.4.14 Cooling Towers Emission Sources

Cooling towers are used extensively in refinery cooling water systems to transfer waste heat from the cooling water to the atmosphere. Atmospheric emissions from the cooling tower consist of fugitive VOCs and gases stripped from the cooling water as the air and water come into contact. These contaminants enter the cooling water system from leaking heat exchangers and condensers.

Although the predominant contaminants in cooling water are VOCs, dissolved gases such as H_2S and ammonia may also be found. Control of cooling tower emissions is accomplished by reducing contamination of cooling water through the proper maintenance of heat exchangers and condensers. The effectiveness of cooling tower controls is highly variable, depending on refinery configuration and existing maintenance practices.

1.4.15 Odors in Refineries

Oil refineries and petrochemical plants may generate large volumes of odor from the processing and refining of different fuels. Sulfides, mercaptans, and hydrocarbon compounds are all closely related to the oil industry. The odors generated from these industrial compounds are of a high nuisance value and generally create concern among local air boards and the public. Some typical refinery smells, their possible source, and the most probable compounds contributing to the odors are summarized as follows in Table 1.8.

Table 1.9 showing the most common odor-causing compounds found in the refinery emission is given here under.

Table 1.8 Refinery odors and sources

Type of smell	Source	Odors compounds
Bad eggs	Crude storage	H_2S + trace of disulfides
Sewer smell	Distillation of gases; sulfur removal, flare stacks (cold flare) effluent water, biological treatment plants; LPG odorizing spent caustic loading and transfer	Dimethyl sulfide, ethyl and methyl mercaptans
Burnt oil	Catalytic cracking unit; coking asphalt blowing; asphalt storage	Unsaturated hydrocarbons
Gasoline	Product storage, CPI and API separators	Hydrocarbons
Aromatic (benzene) hot tar	Aromatic plants; naphtha reformers; asphalt storage	Hydrocarbons mercaptans, H_2S

1.4.16 Petrochemicals

Pollutant from petrochemical plant is given in Table 1.10.

Major air contaminants emitted in fertilizer and other petroleum industries are as follows in Table 1.11.

1.5 Standard Level for Air Quality

1.5.1 Ambient Air Quality Standards

Standard for desirable community levels in terms of concentrations and sample arranging time periods, associated with measurements of these pollutants in community air, is described in Table 1.12.

Certain materials have been designated as hazardous air pollutants, to which no ambient air quality standard is applicable, and for which national emission standards will be used. Various pollutants in different sources that should be considered with emission standard are given in Table 1.13.

1.5.2 Emission Standard

Emission standards may be written in several ways:

1. Process weight rate or mass of emitted pollutant per mass of raw material (e.g., 0.15 kg particulate per metric ton of feed to a kiln).
2. Mass of emitted pollutant per mass of product (e.g., 1.5 kg of nitrogen oxide per metric ton of nitric acid produced).
3. Mass of emitted pollutant per volume of emitted gas (e.g., 90 mg of particulate per dry standard cubic meter).

Table 1.9 Odors of compounds which may be found in refinery emissions

Chemical	Odor chemical threshold, parts per billion by volume (ppb)	Odor description
Acetic acid	1,000	Sour
Acetone	100,000	Chemical, sweet
Amine, monomethyl	21	Fishy, pungent
Amine, dimethyl	47	Fishy
Amine, trimethyl	0.2	Fishy, pungent
Ammonia	46,800	Pungent
Benzene	4,700	Solvent
Benzyl sulfide	2	Sulfidy
Carbon disulfide	210	Vegetable-like, sulfidy
Chlorine	314	Bleach, pungent
Chlorophenol	0.03	Medicinal
Dimethyl sulfide	1–2	Vegetable-like, sulfidy
Diethyl sulfide	6	Garlic-like, foul
Diphenyl sulfide	5	Burnt, rubbery
Hydrogen sulfide	5	Rotten eggs
Methyl ethyl ketone	10,000	Sweet
Mercaptan, methyl	1–2	Sulfidy, decayed cabbage
Mercaptan, ethyl	0.4–1	Sulfidy, decayed cabbage
Mercaptan, n-propyl	0.7	Sulfidy
Mercaptan, n-butyl	0.7	Strong, sulfidy
Paracresol	1	Tarry, pungent
Paraxylene	470	Sweet
Phenol	47	Medicinal
Phosphine	21	Oniony, mustard
Sulfur dioxide	470	Sharp, pungent
Toluene	2,000–4,700	Solvent, moth balls
Butane	6,000	
Heptane	18,000	
Amylenes and pentenes	170–2,100	

Table 1.10 Pollutant from petrochemical plants

Combustion	SOx, NOx, and other pollutants in waste gas from boiler for power generation, heating furnace (fractionator reboiler), and cracking furnaces (naphtha, EDC, steam reforming)
Evaporation and drying	Drying of synthetic rubber, plastics—emission of dry air containing solvent and monomer
	Vaporization from storage tanks (mostly when charging)—emission of hydrocarbons
	Opening of vessel manholes—emission of hydrocarbons
Off-gas, vent gas, flare stack gas	Air used for oxidation, oxychlorination, and ammoxidation changes into a waste gas consisting of nitrogen, carbon dioxide, and a bit of by-products
	Non-condensing gas from distillation tower and reactor is accompanied with hydrocarbons and by-products
	Pollutants in combustion gas from flare stack
Powder handling	Dispersion of plastics during transportation and storage
	Dispersion of catalysts used for fluidized bed
Leakage (loss)	Leakage from pump and compressor shafts and flanges—emission of hydrocarbons

Table 1.11 Major air contaminants emitted in fertilizer and other petroleum industries

Process of operation	Air contaminants emitted
Phosphate fertilizers: crushing grinding and calcining	Particulates (dust)
Hydrolysis of P_2O_3	PH_3, $P_2O_5PO_4$, mist
Acidulation and curing	HF, SiF_4
Granulation	Particulates (dust) (product recovery)
Ammoniation	NH_3. NH_4Cl. SIF_4 HF
Nitric acid acidulation	NO_2 gaseous fluoride compounds
Super phosphate storage and shipping	Particulates (dust)
Ammonium nitrate reactor	NH_3, NO_2
Prilling tower	NH_4, NO_3
NGL units	Hydrocarbon
Oil production units	H_2S odor
Gas compression station	H_2S odor
Gas injection plant	H_2S odor
Sulfur unit	H_2S odor
Oil wells	H_2S odor

Table 1.12 Ambient air quality standards

Parameter	Primary	Secondary
Particulates (μg/m^3)		
Annual geometric mean	75	60
Max. 24-h concentration (not to be exceeded more than once a year)	260	150
Sulfur oxides (μg/m^3)		
Annual arithmetic average	80 (0.03 ppm)	60 (0.02 ppm)
Max. 24-h concentration(not to be exceeded more than once a year)	365 (0.14 ppm)	260 (0.10 ppm)
Max. 3-h concentration(not to be exceeded more than once a year)	–	1,300 (0.50 ppm)
Carbon monoxide (mg/m^3)		
Max. 8-h concentration(not to be exceeded more than once a year)	10 (9 ppm)	10
Max. 1-h concentration (not to be exceeded more than once a year)	40 (35 ppm)	40
Photochemical oxidants (μg/m^3)		
One hour maximum	160 (0.08 ppm)	160
Hydrocarbons (μg/m^3)		
Max. 3-hour concentration 6–9 am (not to be exceeded more than once a year)	160 (0.24 ppm)	160
Nitrogen oxides (μg/m^3)		
Annual arithmetic average, 24-h max. average	100 (0.053 ppm)	100

4. Mass of emitted pollutant per heat input (e.g., 0.34 g of sulfur dioxide per 10 6 J of heat input).
5. Plume opacity (e.g., plume not to exhibit more than 20 % opacity). Other emission standards include the limitation of sulfur content of fuel and the restriction on pollutant mass emitted by new vehicles during a specified dynamometer test cycle.

1.6 Process Control of Atmospheric Emissions

1.6.1 Refineries Source of Atmospheric Emissions

Refinery atmospheric emissions and control procedures are customarily considered by types of equipment employed rather than by refinery process operation.

Table 1.13 Various pollutants in different sources

Source	Pollutant considered
Petroleum refineries	Particulate matter (PM), NOx, total reduced sulfur (TRS), odor, CO, H_2S, hydrocarbons (HC)
Asphalt batch plants	Particulate matter (PM), NOx, odor, hydrocarbons (HC)
Phosphate fertilizer plants	Particulate matter (PM), F
Phosphorus reduction plants	Particulate matter (PM), F
Industrial size fuel-burning plants	PM, SO_2, NOx
Storage vessels for petroleum liquids	Hydrocarbons (HC)
Oil & gas NGL, gas INJ./comp. units & PM, HC, CO, CO_2, H_2S, NOx, smoke	Particulate matter (PM), hydrocarbons (HC), CO, CO_2, H_2S, NOx, smoke
Wellhead separator	
Gas sweetening plant	Particulate matter (PM), hydrocarbons (HC), H_2S, RSH, CO, CO_2, NOx, smoke
Desalting plants	Particulate matter (PM), hydrocarbons (HC), smoke, CO, CO_2, H_2S

1.6.1.1 Storage Tanks

Hydrocarbon vapors may be released through a number of mechanisms in storage tanks, including tank breathing due to temperature changes, direct evaporation, and displacement during filling. The principal source of potential loss is from crude oil and light distillate products. The hydrocarbon content of crude oil is substantially saturated and, as explained above, is not believed to be involved in the photochemical smog complex. Light distillates have considerable value and are normally controlled to a practical economic level. Vapor conservation storage may involve tanks with floating roof covers, pressurized tanks, and connections to vapor recovery systems.

For materials exhibiting a true vapor pressure (TVP) at storage conditions of below 78 mmHg, no controls are required; for materials with a TVP between (78 and 570 mmHg), the storage vessel shall be equipped with a floating roof or equivalent; for materials with a TVP exceeding 570 mmHg, the vessel shall be equipped with a vapor recovery system or equivalent.

1.6.1.2 Catalyst Regeneration Units

Coke formed on the surface of catalysts during catalytic cracking, reforming, and hydrogenation is burned off in regenerating vessels by controlled combustion. Flue gases from regenerators may contain catalyst dust, carbon monoxide, hydrocarbon (principally methane), and sulfur and nitrogen oxides. The catalyst dust may be controlled by mechanical or electrical collecting equipment.

The carbon monoxide and unburned hydrocarbons are generally dispersed in the atmosphere, but may be eliminated by burning in a waste heat boiler which also generates additional steam. Amount of catalyst emitted from fluid catalytic cracking regenerator should not exceed one kilogram of catalyst per 1,000 kg of coke make and carbon monoxide to 500 ppm by volume.

In situ-type operation is finding wider use in fluid catalytic cracking unit.

1.6.1.3 Wastewater Separators

Wastewater gravity separators are commonly used to trap and recover oil discharged to the sewer system from equipment leaks and spills, shutdowns, sampling, process condensate, pump seals, etc. Depending on the quantity and type of oil in the sewers, some hydrocarbon vapors may evaporate from the drainage and separator system. If this vaporization is sizable and control is indicated, the front end of the separators may be covered. Catch basin liquid seals, manhole covers, and good house-keeping practices will likewise control drainage system vapor losses.

1.6.1.4 Loading Facilities

While most petroleum products leave the refinery through pipelines with no emission to the atmosphere, loading into tank trucks, tank cars, and drums can result in hydrocarbon vapor loss by displacement or evaporation. Careful operation to minimize spillage, and vapor collection and recovery equipment, will control vapor loss from this operation.

1.6.1.5 Pipeline Valves

The typical refinery contains a maze of piping, mostly above grade. The effects of heat, pressure, vibration, and corrosion may cause leaks in valved connections. Depending on the product carried and the temperature, the leaks may be liquid, vapor, or both. Regular inspection and prompt maintenance will correct vapor loss from this source.

1.6.1.6 Pumps and Compressors

Hydrocarbons can leak at the contact between the moving shaft and stationary casing in pumps and compressors. Asbestos or other fibers are packed around the shaft to retard leakage from shaft motions. Mechanical seals, consisting of two plates, perpendicular to the shaft, forced tightly together, are also used. Wear can cause both packed and mechanical seals to leak product. Inspection and maintenance, sealing glands under pressure, and use of mechanical seals in light hydrocarbon service are useful control measures.

1.6.1.7 Blowdown System, Flares, and Shutdowns

Refinery process units and equipment are periodically shut down for maintenance and repair. Since these turnarounds generally occur about once a year, losses from this source are sporadic. Hydrocarbons purged during shutdowns and start-ups may be manifolded to blowdown system for recovery, safe venting, or flaring.

Vapors can be recovered in a gas holder or compressor and discharged to the refinery fuel gas system. Flares should be of the smokeless type, utilizing either steam or air injection. Design data for smokeless flares are readily available from combustion equipment manufacturers and in industry technical manuals. For aesthetic purposes, ground flares are becoming more popular.

1.6.1.8 Boilers and Process Heaters

Refineries depend on boilers and heaters to supply high-pressure steam at elevated temperatures. Fuels may include refinery or natural gas, heavy fuel oil, and coke, often in various changing combinations.

Sulfur oxides in the flue gas are, of course, a result of the sulfur in the fuel feed. Nitrogen oxides and small quantities of hydrocarbons, organic acids, and particulate matter are also present. Sulfide stripping of fuel gas prior to burning and selective blending of fuels may be employed to control sulfur emissions. Normally, good combustion practices will control smoke and particulate matter.

Stacks on boilers and heaters are elevated to improve atmospheric dispersion and further diminish resulting ground-level concentrations of gases such as sulfur dioxide and nitrogen oxides.

1.6.1.9 Sulfur Recovery Units

The potential for H_2S loss to the atmosphere occurs at two points in the Claus process:

(a) the interface between the H_2S recovery process and the feed to the unit.
(b) the tail gas from the plant. If the Claus plant breaks down and cannot take the feed, H_2S has to be flared. Some refineries utilize spare Claus capacity so that flaring is minimized.

The presence of H_2S in the tail gas indicates that the conversion of H_2S to S is not complete. This tail gas is either incinerated at 538–649 °C temperature thus releasing all sulfur compounds as sulfur oxides, or further treated in a tail-gas recovery system. The tail-gas treating systems generally convert all sulfur compounds to H_2S, after which the stream is scrubbed to remove the H_2S.

Incinerators

There are a number of incineration processes available for the incineration of refinery sludges, solids, and spent caustics. Fluid bed incinerators can be installed in refineries. These fluid bed incinerators operate at temperatures of 704 °C.

As the sludge is burned, the solids from the sludge remain in the bed, while gaseous products of combustion, water vapor, and fine particulate matter pass overhead through a cyclone separator and a water scrubber before venting to the atmosphere.

1.6.1.10 Miscellaneous

Various other miscellaneous emission sources, usually of lesser significance, will be found in refinery operations. Pressure relief valves may be manifolded into vapor recovery or flare systems to control leakage and relief discharged. Steam-driven vacuum jets, employed to induce negative pressure in process equipment, may discharge light hydrocarbons with the exhaust steam. These gases may be vented and burned in an adjacent boiler or heater firebox.

Fumes from air blowing operations may be consumed by incineration or absorbed by scrubbing. Gases from spent caustic and mercaptan disposal may be burned in fireboxes.

Hydrogen sulfide and mercaptans are the principal potential pollutants that may cause odors. These gases can be released from process steam condensates, drain liquids, barometric condenser sumps, sour volatile product tankage, and spent caustic solutions from treating operations.

Odorous compounds in steam condensates can be removed by stripping with air, flue gas, or steam, and offensive gases can be burned in furnaces or boilers. Drain liquids can be collected in closed storage systems and recycled to the process.

Barometric condensers are being replaced by more modern surface condensers, and the non-condensables may be burned in process heaters or in a separate incinerator.

Spent caustic can be degasified, neutralized with flue gas, and/or stripped before disposal. Sulfides can also be removed from sour process water and spent caustic solutions by air oxidation to thiosulfates and sulfates.

Refinery waste gases that contain hydrogen sulfide are generally scrubbed with appropriate solutions for extraction of the sulfide by non-regenerative or heat-regenerative procedures. In the former method, the waste gases are scrubbed with a caustic solution, producing a solution of sodium sulfide and acid sulfide. As described above, the spent caustic may be oxidized by air blowing or sold. Vent gas from the blowing operation should be burned.

Heat-regenerative methods involve scrubbing sour gases with various types of amine, phenolate, or phosphate solutions which absorb hydrogen sulfide at moderately low temperatures and release it at higher temperatures. These methods are cyclic and consist of an absorption step, in which the hydrogen sulfide is scrubbed from the absorbing solution at approximately 38 °C, followed by a regeneration step, in which the solution is reactivated for use by heating it to its boiling point to drive off the hydrogen sulfide. The released hydrogen sulfide is then burned or oxidized to form sulfur.

Refinery process changes may have the net effect of reducing overall emissions to the atmosphere. Examples of such process changes include substitution of hydrogen treating for chemical treatment of distillates, use of harder catalysts to reduce attrition losses and regeneration of spent chemicals for reuse.

1.6.2 Odor Control Methods for Refiner Purpose

1.6.2.1 Alkylation Units and Asphalt Oxidizers

Odor emissions are effectively prevented by afterburners or the collection and feeding the off-gases to process heaters.

1.6.2.2 Barometric Condensers and Vacuum Towers

Odor prevention has been achieved by venting all off-gases, including vacuum jet gases, to process heaters or where necessary, flares. If the off-gases contain high concentrations of sulfur compounds, further improvement may be obtained by pretreating the off-gases prior to burning. Condensing the vacuum jet stream and adding the condensate to the foul water stripper have also been effective. In many refineries, the barometric condensers are being replaced with closed units which effectively eliminate this source.

1.6.2.3 Bio-Oxidation Units

Odor emissions have been controlled by separating and pretreating particularly odorous streams. Proper maintenance of pH levels and mechanical aeration have also decreased odor emissions. In addition, proper handling of the bio-oxidation unit sludge improves odor prevention.

1.6.2.4 Catalytic Cracking Units

Many odor problems have essentially been eliminated by the replacement of older moving bed units with new fluid catalytic cracking units. In addition, desulfurizing the feedstock has been effective.

1.6.2.5 Caustic Treatment

Odor problems have been diminished by condensing the off-gases, recovering spent caustic by air blowing and burning the off-gases, or converting the thiols and extracting the resulting disulfides.

1.6.2.6 Cokers

Odor emissions are controlled by extracting the off-gas with gas–oil streams or condensing the off-gases and feeding the non-condensables to a process heater unit. The water quench is also a source of odors which can be eliminated by feeding the quench vapors to a flare or condensing it and adding the condensate to the foul water stripper. The use of an amine scrubber on the off-gases with the scrubbed stream added to the fuel gas line is also effective.

1.6.2.7 Desalter

Odors have been prevented by feeding the desalted water to a sour water stripper or in some circumstances by disposal in deep wells.

1.6.2.8 Incinerator

Odor emissions can be controlled by increasing the combustion temperature.

1.6.2.9 Lube Oil

Odors can be significantly reduced through the use of solvent extraction in place of acid treatment. Further improvement has been obtained by adding vapor control systems to the solvent storage tanks, installing blower–scrubber systems to the vents from blending and canning areas, and the use of hydrotreating instead of H_2SO_4 in the production of white oil.

1.6.2.10 Propane Loading

Odor problems are commonly produced during the addition of an odorant. The use of a totally enclosed odorant system and a flush of the piping with the propane after odorant addition eliminate the major source of odor.

1.6.2.11 Pumps and Compressors

Odors are controlled by the replacement of packed seals with mechanical seals. Where mechanical seals cannot be used, the pump seals can be enclosed and vented to flare.

1.6.2.12 Safety Flares

Flares should only be used for upset conditions to minimize odor problems.

1.6.2.13 Sewer Systems

Odor problems may be prevented by adding oil separators to each process unit, pretreating the effluent before discharge to the sewer and flashing the desalted water before it discharged to the sewer. Where possible, all hot spent process water should be cooled before dumping in the sewer.

1.6.2.14 Sour Water Stripping

Odors created can be prevented by collecting the off-gases and sending them to a sulfur recovery unit, or by incineration.

1.6.2.15 Spills and Leaks

Odor problems may be minimized by collecting the spill in vacuum trucks for quick disposal.

1.6.2.16 Steaming of Towers and Vessels

Odor emissions can be reduced by conducting a more complete pump-out and through the use of a water or gasoline wash prior to steaming. Improvement can also be obtained by containing the first steam and venting it to a flare and treating the wastewater.

1.6.2.17 Storage Tanks for Volatile Materials

Storage tank odor prevention has been improved by the use of floating roofs and not pumping below the level where a proper seal is maintained. Odor prevention from asphalt tanks or during cleaning may be improved by venting through a charcoal canister.

1.6.2.18 Sulfur Recovery Unit

Odor prevention may be obtained by maintaining adequate temperatures in the process unit and storing the sulfur in molten form in a closed system.

1.6.2.19 Vapor Collection System

Odors are reduced by recovery of all off-gases and vapor emissions in a completely closed system for use as fuel gas, recycle for reprocessing, or to be sent to a combustion device for incineration.

1.6.2.20 Hydrogen Treating

Odors compounds are treated with hydrogen to produce desirable products and H_2S.

1.6.3 Petrochemical Processes Source of Atmospheric Emissions

Petrochemical process that uses air-oxidation-type reaction normally has large continuous amounts of gas emissions to the atmosphere. The six processes considered in this section employ reactions using air oxidation.

1.6.4 Acrylonitrile

Acrylonitrile is a colorless liquid with a slightly pungent odor. It is used in the manufacturing of resins, nitrile elastomers, and as an intermediate in the production of adiponitrile and acrylamide.

Acrylonitrile is mainly produced by the "Sohio" process. The Sohio process involves the ammoxidation of propylene in the presence of a catalyst at 350–600. The catalysts used are mixed metal oxides such as iron–antimony oxides, uranium–antimony oxides, bismuth–molybdenum oxide. Propylene and ammonia are reacted along with oxygen in the presence of a catalyst. The reactor effluent consists of major products such as acrylonitrile and acetonitrile, waste gases, and unconverted raw material.

The major source of emissions from an acrylonitrile plant is the exit-gas stream from the product absorber. For a typical plant producing 90 million kg a year of acrylonitrile, the gas flow rate would be 1,284 standard cubic meter per minute (scmm). The pollutants in kilograms per metric ton of produced acrylonitrile are shown in Table 1.14.

In Fig. 1.1, flow manufacture of acrylonitrile has been shown. It shows the process flow diagram for the acrylonitrile process.

The composition of the stream will depend primarily on the selectivity of the catalyst in the reactor and the efficiency of the absorber.

A second, less important, source of emissions from this process is the by-product incinerator. About 1–3 % of the nitrogen content of the excess hydrogen

Table 1.14 The pollutants in kilograms per metric ton of produced acrylonitrile

Pollutant	Kilogram per ton of acrylonitrile produced (with no control)
Carbon monoxide	122
Propylene	38
Propane	61
Hydrogen cyanide	0.5
Acrylonitrile	0.25
Acetonitrile	6.5

cyanide and acetonitrile is converted into nitrogen oxides in a commercial incinerator operating at 871 °C. There is presently no demonstrated control device for the oxides of nitrogen. The amount of the emission depends upon the amount of by-product gas burned, the temperature of combustion, and the percent of excess air.

A thermal incinerator will oxidize essentially all of the contaminants from the main process vent stream. However, gas has such a low heat content that supplemental fuel must be added to achieve stable flame control and complete combustion. Various combinations of heat exchange could be used.

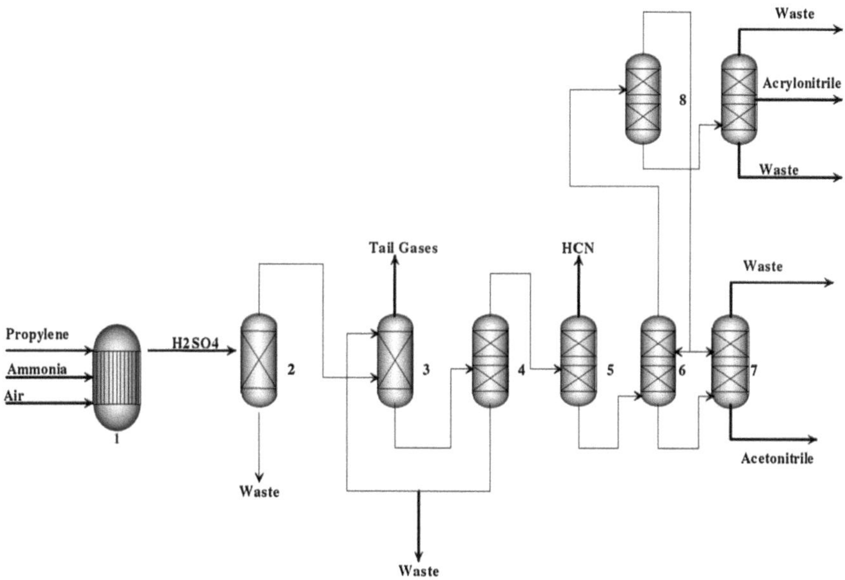

Fig. 1.1 Flow diagram of the Sohio process for the manufacturing of acrylonitrile, *1* CSTR reactor, *2* neutralizer, *3* absorber, *4* stripper, *5* HCN column, *6* extractive distillation column, *7* acetonitrile purification column, *8* acrylonitrile purification columns process flow diagram of the acrylonitrile process (Reprinted with permission © Elsevier, [20] C. Driussi, J. Jansz, Journal of Cleaner Production 14 (2006) 682–688)

Because acrylonitrile plants generate more steam than they can utilize, any steam generated by the incinerator has to be exported. The most feasible air pollution control system for new plants would be a thermal incinerator with process vent gas preheat, combustion air preheat, and a waste heat boiler.

1.6.5 Carbon Black

Of all carbon black produced, about 84 % is manufactured by the furnace process. The thermal process which produces 14 % of the carbon black is a minor source of air pollution. The plants using the channel process presently account for less than 2 % of production.

Figure 1.2 shows that previously heated oil, gas fuel, and high-temperature air are injected into a reactor, creating high-temperature combustion gas. Feedstock oil is atomized into this high-temperature combustion air, producing carbon black. Later, cooling water is atomized to stop carbon black generating reaction.

Inside bag filters, carbon black and by-product gas are separated. At this stage, carbon black is in powder form. It is then mixed with water inside a pelletizer and processed into pellets.

Carbon black stored in product tanks is packed according to customer needs in bulk (trucks or freight trains), flexible containers, etc., and transported. By-product gas separated inside the bag filters is sent to the driers for reuse and to the power plant as fuel.

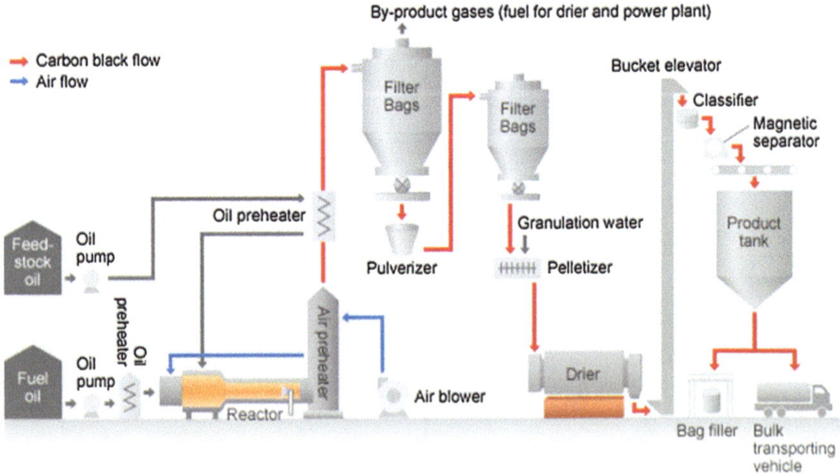

Fig. 1.2 Carbon black process diagram

1.6.6 Furnace Process

In the furnace process, natural gas and high carbon aromatic oil are preheated and introduced into a furnace with a limited amount of air (Fig. 1.3). A combination of cracking and combustion occurs.

The major source of emissions from a furnace black plant is the exit-gas stream from the bag collector. For a typical plant producing 41 million kg of black per year, the gas flow rate would be 1,440 scmm.

The pollutants in pounds per ton of black produced are shown in Table 1.15. The exact composition of the stream will depend upon the grade of black produced and the composition of the aromatic oil.

Grades which require a larger proportion of gas feed will have increased emissions of carbon monoxide. The amount of hydrogen sulfide formed is proportional to the sulfur content of the aromatic oil.

Any type of incineration device will oxidize most of the carbon monoxide and hydrogen sulfide in the vent stream. However, a carbon monoxide boiler or thermal incinerator would be more efficient than a flare. Because off-gas is low in heat content [about 40 BTU per scf (356,000 cal per scm)], most combustion devices require supplemental fuel (natural gas) to maintain combustion.

Various combinations of heat exchange, supplemental fuel addition, and heat recovery can be used. The choice is an economic one determined by the costs of fuel and equipment and the ability to use any steam generated. The most efficient air pollution control system for new plants under most circumstances would be a process vent gas thermal incinerator with combustion air heat exchange plus a waste heat boiler and steam-driven process equipment. This should oxidize all the carbon monoxide, hydrogen sulfide, and hydrocarbons.

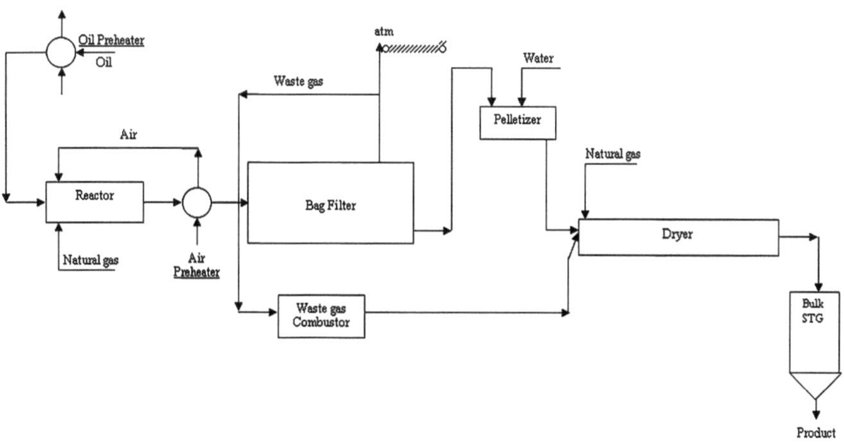

Fig. 1.3 Furnace process

Table 1.15 Pollutant production in the furnace black process

Pollutant	Kilogram per ton of carbon black produced
Hydrogen	116.5
Carbon monoxide	1,270
Hydrogen sulfide	27.5
Sulfur dioxide	Trace
Methane and acetylene	54.5
Particulate matter (black)	2

1.6.7 Ethylene Dichloride

Ethylene dichloride is produced by the direct chlorination or oxychlorination of ethylene. The atmospheric pollutants from direct chlorination processes are much less than those from the oxychlorination process. Almost all ethylene dichloride is used for the manufacturing of vinyl chloride monomer.

1.6.7.1 Oxychlorination Process

Vaporized ethylene, anhydrous hydrogen chloride, and air are fed to a catalytic reactor which operates at 2.3–6.1 atm and 200–315 °C. The reaction is

$$2C_2H_4 + O_2 + 4HCl = 2C_2H_4Cl_2 + 2H_2O \tag{1.4}$$

1.6.7.2 Air Pollutant Emissions and Their Control

The major source of emissions from an ethylene dichloride plant is the exit-gas stream from the solvent scrubber (see Figs. 1.4, 1.5).

For a typical plant of the type described producing 317 million kg per year, the gas flow rate would be 489 scmm.

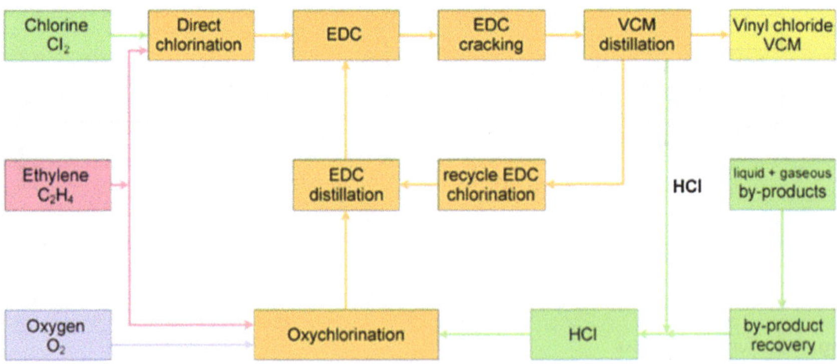

Fig. 1.4 A schematic to manufacture vinyl chloride monomer

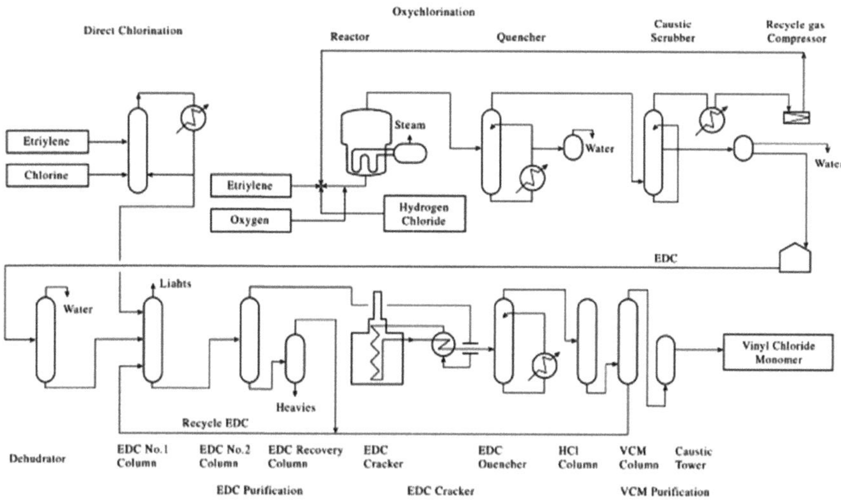

Fig. 1.5 A process flow diagram to manufacture vinyl chloride monomer (Reprinted with permission © Elsevier, Y. Saeki, T. Emura/Prog. Polym. Sci. 27 (2002) 2055–2131)

The pollutants in (kilograms per metric ton) of ethylene dichloride produced are shown in Table 1.16. The values given are average values for several different oxychlorination processes. The composition of the stream will depend on catalyst activity, reactor operating conditions, and the solvent scrubber efficiency. All existing ethylene dichloride plants vent this stream directly to the atmosphere with no treatment. Any one of several combustion devices could be used to destroy the contaminants present.

A water or caustic scrubber could be used after the combustor to remove the hydrogen chloride generated. Several different types of incineration would be feasible. The most efficient method for new plants would be a thermal incinerator, followed by a waste heat boiler and a final caustic scrubber.

Table 1.16 Pollutant production in ethylene dichloride manufacturing by oxychlorination

Pollutant	Kilogram per ton of ethylene dichloride produced
Carbon monoxide	0.65
Methane	2
Ethylene	4.8
Ethane	6.3
Ethylene dichloride	6.9
Ethyl chloride	5.9
Aromatic solvent	1.1

The steam generated could be used in the plant. Because this system has not been demonstrated, it is possible that there might be significant operating problems. Careful control of temperature would be necessary to prevent condensation of hydrogen chloride and resultant corrosion of metal. Also, because of the low heating value of the gas stream 267,000–445,000 cal per sccm, supplemental fuel would have to be added to achieve complete combustion and satisfactory flame control.

The most feasible air pollution control system for an existing plant would be a thermal incinerator and scrubber on the main process vent.

1.6.8 Ethylene Oxide

The oxidation of ethylene is the most widely used process for the production of ethylene oxide. The production of ethylene glycol, used primarily as automotive antifreeze, consumes more than half of the ethylene oxide produced. The second largest use of ethylene oxide is in the manufacturing of non-ionic surfactants.

1.6.9 Oxidation of Ethylene

Ethylene oxide is produced by passing ethylene and air (or oxygen) over a silver catalyst and recovering ethylene oxide from the gas stream by water absorption. Ethylene oxide is stripped from the water solution and refined. The reaction is

$$C_2H_4 + {}^1/_2O_2 \rightarrow C_2H_4O + Heat \tag{1.5}$$

Flow diagrams for the air and oxygen processes are given in Figs. 1.6 and 1.7. In the air process, air and ethylene are added to a recycle gas stream and fed to the main reactor where the reaction takes place in tubes containing silver catalyst.

After heat exchange with recycle gas, the reactor effluent gases pass to the main absorber where the ethylene oxide is absorbed in a water solution. Approximately two-thirds of the absorber overhead gas (which contains nitrogen, carbon dioxide, unreacted ethylene, and air) is returned as recycles to the main reactor. The remaining one-third of the main absorber overhead gas is purged from the system. Most of the ethylene in the purge stream is recovered as ethylene oxide which is formed in the purge reactor and removed in the purge absorber.

The vent from the purge absorber, mostly nitrogen and carbon dioxide, contains some unreacted ethylene, a small amount of ethylene oxide, and all the ethane in the ethylene raw material. The ethylene oxide in the main absorber and purge absorber bottoms is removed by desorption, refined, and sent to storage.

Although the oxygen process is similar to the air oxidation process, there is usually no purge reactor or absorber. Also, because the conversion of ethylene per pass is lower, the recycle is larger. Product is recovered by absorption as in the air oxidation process.

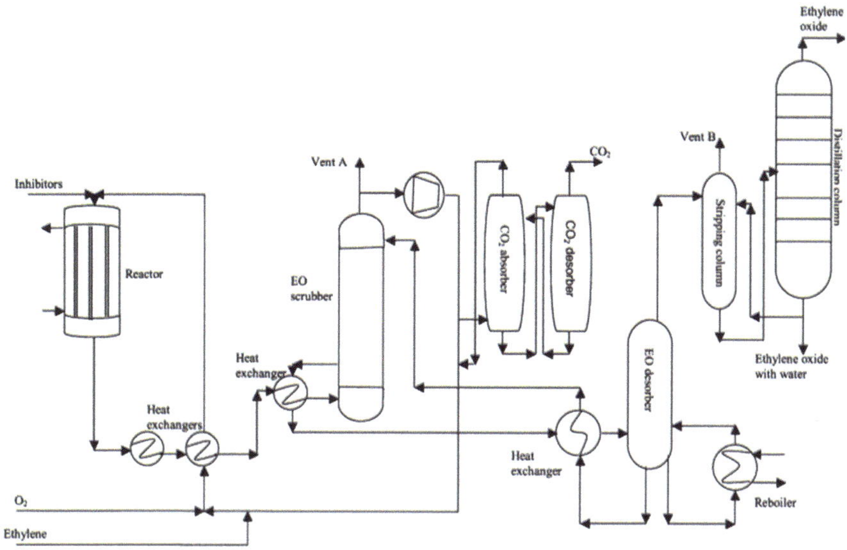

Fig. 1.6 A process flow diagram to produce ethylene oxide (Reproduce with Permission © Elsevier, [40] F.I. Khan et al./Journal of Loss Prevention in the Process Industries 15 (2002) 129–146)

A carbon dioxide absorber, used on a portion of the recycle stream, controls the buildup of carbon dioxide. Other inert gases are removed by a small absorber vent gas purge. In a variation of the process, methane is added to the recycle gas to act as an inert in the reactor and yield a high calorie vent gas which burns easily in a boiler.

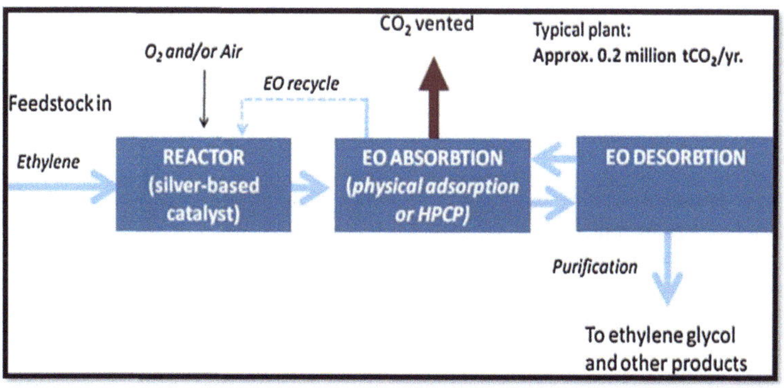

Fig. 1.7 Block flow diagram of the direct oxidation process for the manufacturing of ethylene oxide using air, noting potential sources of air pollution

1.6.10 Air Pollutant Emissions and Their Control

The only important source of emissions from ethylene oxide manufacture is the purge vents. Typical pollutants for the purge absorber vent of an ethylene oxide plant using air oxidation are shown in Table 1.17. For a typical 90 million kg per year plant, the vent gas rate would be 9,775 sccm. Typical main process vent pollutants for a plant using oxygen oxidation are shown in Table 1.18. For a 90 million kg per year plant, the main process vent gas rate would be 3.4 sccm. Because the carbon dioxide vent of 164 sccm contains an insignificant amount of contaminants, it can be vented or sold as carbon dioxide.

The most efficient method to control emissions from an ethylene oxide plant is to feed the main process vent to a catalytic combustor and use the hot gases from this unit in expanders to drive the process compressors. More than 99 % removal of combustible gases would be expected.

1.6.11 Formaldehyde

All formaldehyde was produced by the air oxidation of methanol. About 57 % was used in combination with phenol, urea, or melamine to form resins and adhesives. The largest and fastest growing use of phenol–formaldehyde resins is as an adhesive for plywood. Urea formaldehyde resin is primarily used as an adhesive for particle board. Formaldehyde is also used to manufacture hexamethylenetetramine, pentaerythritol, and several miscellaneous resins. Figure 1.8 shows a flow diagram for manufacturing formaldehyde.

1.6.11.1 Air Pollutant Emissions and Their Control

The major source of emissions from a formaldehyde plant is the exit-gas stream from the scrubber. The amount and composition of gaseous emissions and the type of control device that can be used depend upon the type of process.

For the mixed metal oxide catalyst process, the pollutants in points per ton of 37 % formaldehyde solution produced are shown in Table 1.19. The values given are for a plant operating at maximum recycle; for non-recycle operation, these

Table 1.17 Pollutant production in ethylene oxide manufacturing by air oxidation (kilograms per metric ton)

Pollutant	With no control	Kilograms per metric ton
Ethane	6	0
Ethylene	92	0
Ethylene oxide	1	0

Table 1.18 Pollutant production in ethylene oxide manufacturing by oxygen oxidation (kilograms per metric ton)

Pollutant	With no control	With combustion
Ethane	3	0
Ethylene oxide	2.5	0

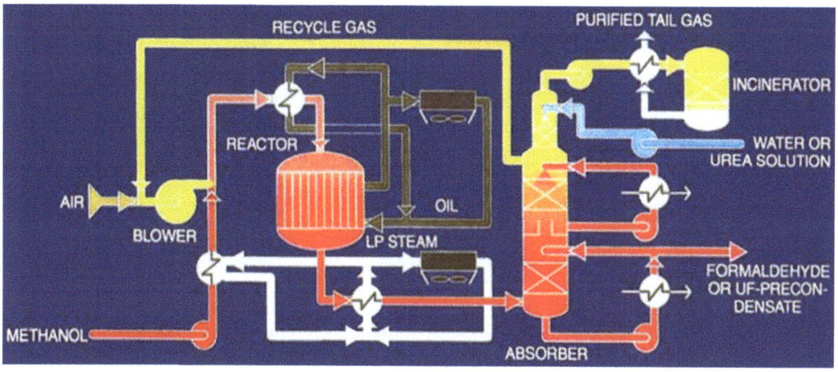

Fig. 1.8 Flow diagram of the methanol process for the manufacture of formaldehyde, noting potential sources of air pollution

Table 1.19 Maximum absorber vent gas recycle pollutants in formaldehyde production by mixed oxide catalyst process (values are kilograms per metric ton)

Chemical component	With no control	With incineration	With water scrubbing
Formaldehyde	0.8	0	0.1
Methanol	2	0	0.2
Carbon monoxide	17	0	17
Dimethyl ether	0.8	0	0.8

values would be higher. For a typical plant producing 45 million kg per year of 37 % formaldehyde, the exit-gas flow rate at maximum recycle would be 9.6 sccm. Water scrubber could be used to remove formaldehyde.

An incinerator is used to oxidize all contaminants from this stream, but because of the very low heat content of the gas, considerable auxiliary fuel should be required.

For the silver catalyst process, the pollutants in pounds per ton of 37 % formaldehyde solution produced are shown in Table 1.20. For a typical plant producing 100 million 1b (45 million kg) per year of 37 % formaldehyde, the gas flow rate would be 62 sccm with a heat content of about 552,000 cal/sccm. The composition of the vent gas depends again on design and operating conditions of the absorber, but the absorption is easier because of the lower gas volume. Incineration can be used to oxidize all the contaminants in this stream.

1.6.12 Phthalic Anhydride

All phthalic anhydride is produced by the vapor-phase air oxidation of either o-xylene or naphthalene. Figure 1.9 shows a flow diagram of phthalic anhydride production using o-xylene.

Table 1.20 Pollutant production in formaldehyde manufacturing by silver catalyst process (values are kilograms per metric ton)

Pollutant	With no control	With combustion
Formaldehyde	0.6	0
Methanol	2.5	0
Carbon monoxide	5.2	0
Hydrogen	10	0

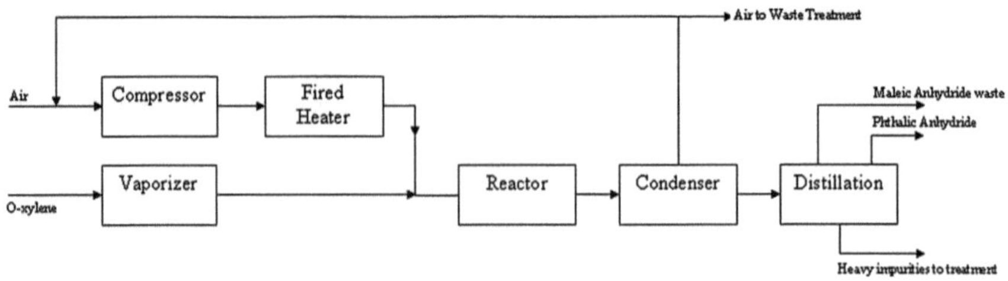

Fig. 1.9 Flow diagram of the oxidation o-xylene or naphthalene for the manufacturing of phthalic anhydride, noting potential sources of air pollution

About 56 % goes into the production of plasticizers for vinyl resins, and phthalic-based alkyd resins, used primarily for exterior surface coatings, account for about 22 % of the material produced. About 17 % is used in the manufacturing of polyester resins, most of which are glass reinforced.

The major source of emissions from phthalic anhydride manufacture is the off-gas from the switch condensers.

For a typical (58 million kg per year) plant, this stream has a flow rate of 3,680 scmm. The organic acids and anhydrides are malodorous lachrymators.

The most common type of emission control is water scrubbing. A typical combination for this application is a Venturi scrubber, followed by a cyclone separator and a packed counter-current scrubber. The absorption efficiencies of the various scrubbers will vary considerably, but typical emissions are given in Table 1.21. The scrubbing liquid must be incinerated in a small natural gas incinerator.

Table 1.21 Pollutant production in phthalic anhydride manufacture (values are kilograms per metric ton)

Pollutant	With no control	With incineration	With water scrubbing
Organic acids and anhydrides	65	2.5	3
Sulfur dioxide	5	5	5
Carbon monoxide	140	0.1	140
Particulate matter	Trace	–	Trace

1.6.12.1 Emission Reduction During air Pollution Episodes

Atmospheric emissions from the petrochemical processes described in this section can be reduced significantly or stopped completely within a period of 1–3 h. Petrochemical plants usually have several independent trains of processing equipment, and these trains may be shut down as required to reduce emissions.

Most operators also have the option of reducing production by leaving all trains on line and reducing the throughout per train. In either case, the emissions from the process would be approximately proportional to operating level.

1.6.13 Fertilizer Industries

Table 1.22 shows major pollutants in fertilizer industries.

1.6.14 Crude Oil Terminals Air Pollutants

Several types of crude terminals are encountered throughout the industry based on the sources and final destinations of the crude oil. The types of terminals to be considered in the following discussion are as follows: inland pipeline terminals, marine shipping terminals, onshore marine receiving terminals, offshore marine receiving terminals, barge shipping terminals, and barge receiving terminals.

A large amount of storage capacity is required at crude oil terminals, not only to enable the crude to be brought into the terminal from numerous producing regions (which may include several different quality crudes), but also to provide the

Table 1.22 Major pollutants in fertilizer industries

Pollutant consider	Process of operation	Control system
Dust	Phosphate fertilizers: crushing, grinding, and calcining	Exhaust system, scrubber, cyclone, baghouse
PH_3, P_2O_5, PO_4 mist	Hydrolysis of P_2O_5	Scrubbers, flare
	Acidulation and curing	Scrubbers
HF, SiF_4 particulates	Granulation	Exhaust system, scrubber, cyclone, electrostatic
	Ammoniation	
NH_3, NH_4Cl, HF, SiF_4	Precipitator, baghouse, high-energy scrubber	
NO_2 gases particulates	Nitric acid acidulation	Scrubber, addition of urea
	Superphosphate storage and shipping	Exhaust system, cyclone, or baghouse
NH_3, NO_2	Ammonium nitrate reactor	Scrubber
NH_4, NO_3	Prilling tower	Proper operation control, scrubbers

tankage for segregation, batching, blending, and inventorying necessary for continuous pipeline operations before the crude oil can be moved to refineries.

Similar operations are necessary for marine terminals both for shipping and for receiving of crude oil.

The sources of pollutants are based on the type of operation (receiving or shipping) and the type of transportation used (pipeline, tanker, or barge).

The most significant air pollutants emitted by a terminal operation are hydrocarbon, which occur at all types of crude facilities. Other air pollutants including odors, SOx, NOx, CO, and particulates are also concerned.

1.6.15 Hydrocarbons

Emission sources include storage tanks (normally onshore), transportation vessel tanks, tanker refueling, tank cleaning (both storage and vessels), tank degassing (preparation for inspection and maintenance), ballasting, and fugitive emissions. In the case of onshore or permanent facilities, the mechanisms for limiting hydrocarbon emissions or collecting the vapors for incineration are similar to those discussed for refineries.

The emissions from operations such as tanker refueling and tank degassing may incorporate vapor recovery or vapor collection/disposal (flaring) systems depending on whether an onshore or offshore facility is involved. Several operational control techniques have been considered as alternatives to the use of vapor control systems.

These approaches include segregated ballasting, tank cleaning, slow loading, short loading, and the routing of vapors into tanks that are being emptied. The hydrocarbon emissions from ballasting on the other hand are normally uncontrolled. Hydrocarbon vapors can also be created by crude oil washing of cargo tanks during unloading.

1.6.15.1 Emissions from Tankers Ballasted Tanks

The emission of hydrocarbon gas from ballasted tanks can be avoided by the use of permanent ballast tanks of sufficient capacity to provide the minimum departure drought (International Marine Organization standards should be used). For more information, please refer to environmental management practices in oil refineries and terminals United Nation Environment Programme 1987.

1.6.15.2 Sulfur Oxides (SOx)

Emissions result from the flaring of H_2S containing gases, space heating with fuel oil, fuel combustion, and tankers. The flaring emissions are uncontrolled, while the fuel combustion emissions may be "controlled" by using the low-sulfur fuel oil.

1.6.15.3 Nitrogen Oxides (NOx)

Sources include space heating, the flare, and tankers. Emission controls are normally not employed due to the small quantities emitted.

1.6.15.4 Particulates

Emissions at the terminal should be limited to fugitive emissions such as dust which can be controlled by methods such as paving or vegetation cover.

1.6.15.5 Odors

Odors from a crude terminal are usually caused by H_2S, mercaptans, and hydrocarbon emissions. The most effective means of controlling odor-causing emissions, especially those containing H_2S, is vapor collection and incineration.

1.7 Classification of Common Gases in Chemical Industries

Table 1.23 lists the gases that are most commonly encountered in chemical industries. Each of those gases is discussed in further detail within this section.

1.7.1 Threshold Limit Values

Threshold limit values (TLV) of airborne substances refer to those concentrations within which personnel may be exposed without known adverse effects to their health or safety. The guideline TLVs given on Table 1.23 have been arrived at through a combination of industrial experience and from both animal and human studies. They are based primarily on recommendations of the American Conference of Governmental Industrial Hygienists (ACGIH) and the US National Institute for Occupational Safety and Health (NIOSH). The values quoted should be regarded as guidelines rather than clear demarcations between safe and dangerous concentrations.

There are two reasons for this. First, wide variations in personal response to given substances occur between individuals, depending upon one's state of health, exposure history, and personal habits (e.g., smoking and use of alcohol). Secondly, TLVs are necessarily based on current scientific knowledge and are subject to revision as new evidence becomes available.

To ensure compliance with statutory requirements, ventilation/environmental engineers and industrial hygienists should familiarize themselves with any mandatory TLVs that have been established within their particular country or state.

Three types of TLVs are expressed in Table 1.23. The time-weighted average (TWA) is the average concentration to which nearly all workers may be exposed over an 8-h shift and a 40-h workweek without known adverse effects. However, many substances are sufficiently toxic that short-term exposures at higher concentrations may prove harmful or even fatal.

Table 1.23 List of the gases that are most commonly encountered in chemical industries

Name	Symbol	Molecular weight	Density relative to dry air	Smell, color, taste	Hazards	Guideline TLVs	Methods of detection (in addition to the methods listed, most of these gases can be detected by gas chromatography and stain tubes)	Flammability limits in air percent
Methane	CH_4	16.04	0.554	None	explosive; layering	1 %: isolate electricity 2 %: remove personnel	Catalytic oxidation, thermal conductivity, optical, acoustic, flame lamp	5–15
Carbon dioxide	CO_2	44	1.519	Slight acid taste and smell	Promotes increased rate of respiration	TWA = 0.5 % STEL = 3.0 %	Optical, infrared	–
Carbon monoxide	CO	28.01	0.967	None	Highly toxic; explosive	TWA = 0.005 % STEL = 0.04 %	Electrochemical, catalytic oxidation, semiconductor, infrared	12.5–74.2
Sulfur dioxide	SO_2	64.06	2.212	Acid taste, suffocating smell	Very toxic; irritant to eyes, throat and lungs	TWA = 2 ppm STEL = 5 ppm	Electrochemical, infrared	–
Nitric oxide	NO	30.01	1.036	Irritant to eyes, nose, and throat	Oxidizes rapidly to NO_2	TWA = 25 ppm	Electrochemical, infrared	–
Nitrous oxide	N_2O	44.01	1.519	Sweet smell	Narcotic (laughing gas)	TWA = 50 ppm	Electrochemical	

(continued)

Table 1.23 (continued)

Name	Symbol	Molecular weight	Density relative to dry air	Smell, color, taste	Hazards	Guideline TLVs	Methods of detection (in addition to the methods listed, most of these gases can be detected by gas chromatography and stain tubes)	Flammability limits in air percent
Nitrogen dioxide	NO_2	46.01	1.588	Reddish brown, acidic smell and taste	Very toxic; throat and lung irritant; pulmonary infections	TWA = 3 ppm Ceiling: 5 ppm	Electrochemical, infrared	
Hydrogen sulfide	H_2S	34.08	1.177	Odor of bad eggs	Highly toxic; irritant to eyes and respiratory tracts; explosive	TWA = 10 ppm STEL = 15 ppm	electrochemical, semiconductor	4.3–45.5
Hydrogen	H_2	2.016	0.0696	None	Highly explosive		Catalytic oxidation	4–74.2
Radon	Rn	222	7.66	None	Radioactive; decays to radioactive particles	1 WL- and 4 WL-months per year	Radiation detectors	–
Water vapor	H_2O	18.016	0.622	None	Affects climatic environment	–	Psychrometers, dielectric effects	–

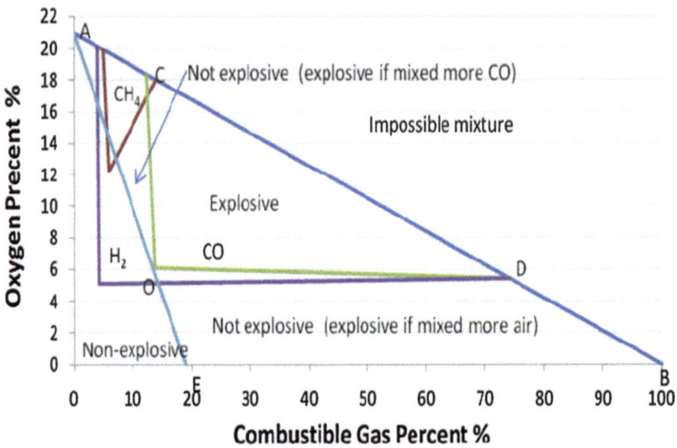

Fig. 1.10 The Coward explosive *triangles* for methane, carbon monoxide, and hydrogen (Reprinted with permission © Elsevier, [50] J. Cheng, Y. Luo/Modified explosive diagram for determining gas-mixture explosibility, Journal of Loss Prevention in the Process Industries, in press (2013) http://dx.doi.org/10.1016/j.jlp.2013.02.007

The short-term exposure limit (STEL) is a TWA concentration occurring over a period of not more than 15 min. That is, concentrations above the TWA and up to the STEL should not last for longer than 15 min. It is also recommended that such circumstances should not occur more than four times per day nor at intervening intervals of less than 1 h. The ceiling limit is the concentration that should not be exceeded at any time. This is relevant for the most toxic substances or those that produce an immediate irritant effect.

1.7.2 Methane, CH_4

Methane is one of the most common strata gases. It is not toxic but is particularly dangerous because it is flammable and can form an explosive mixture with air. This has resulted in the deaths of many thousands of miners. A methane: air mixture is sometimes referred to as firedamp.

Methane itself has no odor but is often accompanied by traces of heavier hydrocarbon gases in the paraffin series that do have a characteristic oily smell. As indicated in Table 1.23, methane has a density that is a little over half that of air. Methane burns in air with a pale blue flame. This can be observed over the lowered flame of a safety lamp at concentrations as small as 1¼ %. In an abundant supply of air, the gas burns to produce water vapor and carbon dioxide.

The explosive range for methane in air is normally quoted as 5–15 % by volume, with the most explosive (stochastic) mixture occurring at about 9.8 %. While the lower limit remains fairly constant, the upper explosive limit reduces as the oxygen content of the air falls.

The flame will propagate through the mixture, while it remains within the flammable range. Figures 1.10 illustrates a well-known diagram first produced by H. F. Coward in 1928 for methane, carbon monoxide, and hydrogen, and Fig. 1.11 shows Cowad diagram for methane in the presence of oxygen. This can be used to track the flammability of air:methane mixtures as the composition varies.

1.7.3 Carbon Dioxide, CO_2

Carbon dioxide appears in subsurface openings from a variety of sources including strata emissions, oxidation of carbonaceous materials, internal combustion engines, blasting, fires, explosions, and respiration.

In addition to diluting oxygen in the air, carbon dioxide acts as a stimulant to the respiratory and central nervous systems. The solubility of carbon dioxide is about 20 times that of oxygen. Diffusion of the gas into the bloodstream is rapid, and the effects on rate and depth of breathing are soon noticed. Cylinders of oxygen used for resuscitation often contain some 4 % of carbon dioxide to act as a respiratory stimulant (carbogen gas). The physiological effects of carbon dioxide have been listed in the literature as follows in Table 1.24.

Fortunately, the administration of oxygen accompanied by warmth and an avoidance of exertion will usually lead to recovery with no known long-term effects.

1.7.4 Carbon Monoxide, CO

The high toxicity of carbon monoxide coupled with its lack of smell, taste, or color makes this one of the most dangerous gases. Carbon monoxide burns with a blue

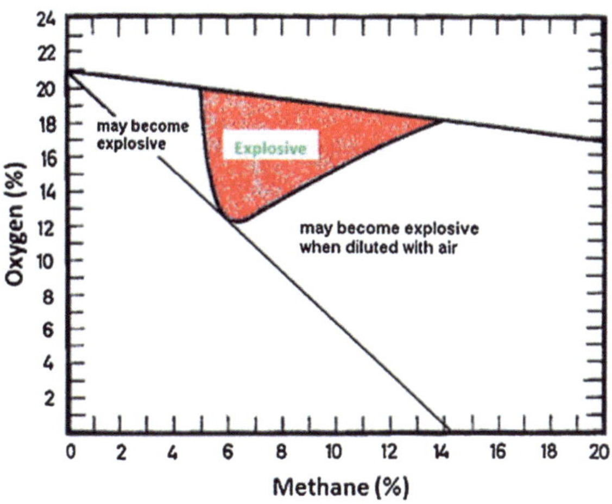

Fig. 1.11 Methane explosibility diagram (Reprinted with permission © Elsevier [60] C.Ö. Karacan et al./International Journal of Coal Geology 86 (2011) 121–156)

Table 1.24 The physiological effects of carbon dioxide

Percent carbon dioxide in air	Effects
0.037–0.038 (The percentage of carbon dioxide in air has been increasing over the past century, currently at a rate of some 1.5 parts per million (0.00015 %) each year. The rise is superimposed on an annual cyclic variation of about 5 ppm (0.0005 %) caused by seasonal reactions of plant life)	None, normal concentration of carbon dioxide in air
0.5	Lung ventilation increased by 5 %
2	Lung ventilation increased by 50 %
3	Lung ventilation doubled, panting on exertion
5–10	Violent panting leading to fatigue from exhaustion, headache
10–15	Intolerable panting, severe headache, rapid exhaustion, and collapse

flame and is highly flammable, having a wide range of flammability, 12.5–74.2 % in air, with the maximum explosibility at 29 %.

It has a density very close to that of air and mixes readily into an airstream unless it has been heated by involvement in a fire, in which case it may layer with smoke along the roof. Carbon monoxide is a product of the incomplete combustion of carbonaceous material.

Carbon monoxide is formed by internal combustion engines, blasting, and spontaneous combustion in coal mines. It can be generated as a component of water gas (carbon monoxide and hydrogen) when water is applied to the base of incandescent coal during firefighting operations.

Unfortunately, hemoglobin has an affinity for carbon monoxide that is about 300 times greater than that for oxygen. To compound the problem, the new substance formed in the bloodstream, carboxyhemoglobin (CO.Hb), is relatively stable and does not readily decompose. The consequences are that very small concentrations of carbon monoxide cause the formation of carboxyhemoglobin which accumulates within the bloodstream. This leaves a reduced number of red cells to carry oxygen molecules throughout the body. The physiological symptoms of carbon monoxide arise because of oxygen starvation to vital organs, particularly the brain and heart.

Physiological reactions to carbon monoxide depend upon the concentration of the gas, the time of exposure, and the rate of lung ventilation, the latter being governed primarily by physical activity. In order to relate the symptoms of carbon monoxide poisoning to a single parameter, the degree of saturation of the blood by carboxyhemoglobin is employed. Although variations exist between individuals, Fig. 1.12 provides a guideline of the progressive symptoms.

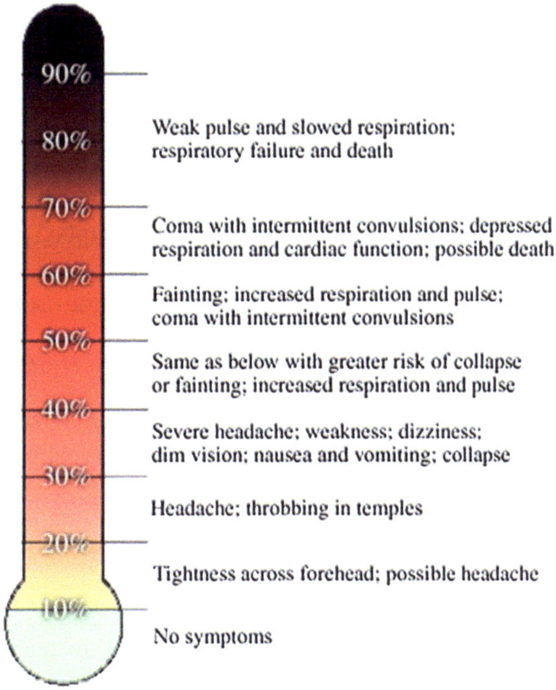

Fig. 1.12 Guideline of the CO progressive symptoms

Persons suffering from carbon monoxide poisoning should be kept warm and removed from the polluted atmosphere, preferably on a stretcher. It may take more than 24 h for blood saturation levels to return to normal. During this time, severe headaches may be experienced.

However, the return to normal blood saturation levels can be accelerated significantly by the administration of pure oxygen. The rapidity with which carbon monoxide is absorbed into the bloodstream and the slowness of its expurgation can result in dangerous blood saturation levels occurring in firefighters who make short repeated expeditions into a polluted area.

In addition to the physical symptoms that have been listed for carbon monoxide poisoning, experience of personnel involved in mine fires has indicated significant psychological reactions that have had grave repercussions. Low levels of blood saturation can give an appearance of intoxication including impairment of judgment and an unsteady gait preceding collapse.

Victims may become silent and morose and may resist or fail to comprehend instructions that will lead them to safety. The sense of time may be affected, a particularly significant symptom when self-rescuer devices are being worn. However, all of these reactions vary considerably between individuals. In particular, total collapse will occur rapidly in high concentrations of carbon monoxide.

A small degree of acclimatization has been observed in persons who are repeatedly exposed to low levels of carbon monoxide as experienced, for example, by habitual smokers. This is thought to occur because of an increase in the number of red cells within the bloodstream.

1.7.5 Sulfur Dioxide, SO_2

Sulfur dioxide (SO_2) is highly toxic gas but one which, fortunately, can be detected at very low concentrations both by its acidic taste and the intense burning sensation it causes to the eyes and respiratory tracts. The latter is a result of the high solubility of the gas in water to form sulfurous acid:

$$H_2O + SO_2 \rightarrow H_2SO_3 \tag{1.6}$$

This, in turn, can oxidize to sulfuric acid, H_2SO_4.

When discussing very small concentrations, it is convenient to refer to parts per million (ppm) rather than percentages. The conversion between the two is accomplished simply by moving the decimal point four places in the appropriate direction. Table 1.25 shows list of physiological reactions to sulfur dioxide that employs parts per million.

First aid for sulfur dioxide poisoning includes the administration of oxygen, immobility, and warmth. The longer-term treatment is for acid corrosion of the eyes and respiratory system.

1.7.5.1 Control Techniques for SOx Emissions Reduction

On the contrary to NOx, SOx emissions are directly linked to the initial sulfur content of the fuel and the combustion parameters do not influent on the amount of SOx emitted. Two strategies can be used to reduce SOx emissions: the formation prevention (low-sulfur fuel usage, fuel desulfurization) or the flue gas desulfurization (wet or dry scrubbing, dual alkali, spray drying, Wellman–Lord process, etc.). There are many post-combustion flue gas desulfurization techniques.

Almost all techniques are based on the acid–alkaline reaction between SO_2 (and SO_3) and an alkaline agent such as lime or limestone, caustic soda, magnesium hydroxide, or ammonia. Other techniques are selective adsorption or absorption. Flue gas desulfurization is mostly used in thermal power plant. Few refineries have

Table 1.25 List of physiological reactions to sulfur dioxide

Concentration of sulfur dioxide ppm	Effects
1	Acidic taste
3	Detectable by odor
20	Irritation of eyes and respiratory system
50	Severe burning sensation in eyes, nose, and throat
400	Immediately dangerous to life

a flue gas desulfurization, except in Japan where principally dry processes are used. The principles of four major techniques are given below.

1.7.5.2 Lime and Limestone Process

Lime and limestone scrubbings are non-regenerative wet processes producing gypsum. Lime and limestone scrubbing are very similar. The use of lime (CaO) instead of limestone ($CaCO_3$) for the slurry preparation is the only difference. The alkaline slurry is sprayed in the absorber and reacts with the SO_2 in the flue gas. The following chemical reactions occur:

SO_2 dissociation:

$$SO_{2(gaseous)} \rightarrow SO_{2(aqueous)} \tag{1.7}$$

$$SO_2 + H_2O \rightarrow H_2SO_3 \tag{1.8}$$

$$H_2SO_3 \rightarrow H^+ + HSO_3^- \rightarrow 2H^+ + SO_3^- \tag{1.9}$$

Lime or limestone dissolution:

$$CaO_{(solid)} + H_2O \rightarrow Ca(OH)_{2(aqueous)} \rightarrow Ca^{2+} + 2HO^- \tag{1.10}$$

Or

$$CaCO_{3(solid)} + H_2O \rightarrow Ca^{2+} + HCO_3^- + HO^- \tag{1.11}$$

Reaction between ions:

$$Ca^{2+} + SO_3^{2-} + 2H^+ + 2HO^- \rightarrow CaSO_{3(solid)} + 2H_2O \tag{1.12}$$

The following reactions can occur if there is excess oxygen:

$$SO_3^{2-} + {}^1\!/_2 O_2 \rightarrow SO_4^{2-} \tag{1.13}$$

$$SO_4^{2-} + Ca^{2+} \rightarrow CaSO_{4(solid)} \tag{1.14}$$

Lime and limestone processes are the most popular flue gas desulfurization system for utility boilers. Some system has achieved SO_2 removal efficiency greater than 95 %. Another advantage is that these processes produce gypsum, which is saleable. However, these processes have limited usage in refineries.

1.7.5.3 Dual-Alkali Scrubbing

Dual-alkali scrubbing is a non-regenerative process using sodium-based solution and lime or limestone to remove SO_2 from flue gases. The following chemical reactions occur:

Main absorption reactions:

$$2NaOH + SO_2 \rightarrow Na_2SO_3 + H_2O \quad (1.15)$$

$$NaOH + SO_2 \rightarrow NaHSO_3 \quad (1.16)$$

$$Na_2CO_3 + SO_2 + H_2O \rightarrow 2NaHSO_3 \quad (1.17)$$

$$Na_2CO_3 + SO_2 \rightarrow Na_2SO_3 + CO_2 \quad (1.18)$$

$$Na_2SO_3 + SO_2 + H_2O \rightarrow 2NaHSO_3 \quad (1.19)$$

$$2NaOH + SO_3 \rightarrow Na_2SO_4 + H_2O \quad (1.20)$$

$$2Na_2SO_3 + O_2 \rightarrow 2Na_2SO_4 \quad (1.21)$$

Regeneration:

$$2NaHSO_3 + Ca(OH)_2 \rightarrow Na_2SO_3 + CaSO_3.\ {}^1\!/_2H_2O \downarrow + {}^3\!/_2H_2O \quad (1.22)$$

$$Na_2SO_3 + Ca(OH)_2 \rightarrow {}^1\!/_2H_2O \rightarrow 2NaOH + CaSO_3.\ {}^1\!/_2H_2O \downarrow \quad (1.23)$$

$$Na_2SO_4 + Ca(OH)_2 \rightarrow 2NaOH + CaSO_4 \downarrow \quad (1.24)$$

This method is attractive because it has a high SO_2 removal efficiency and it reduces scaling problems.

1.7.5.4 Activated Char Process

Activated char process is the principal dry process used in refineries. The circulating activated char absorbs SO_2 at a temperature comprised between 100 and 200 °C. This process has the advantage to also eliminate NOx present in the flue gases. The following chemical reactions occur:

Absorption on char and conversion into sulfuric acid:

$$SO_2 + {}^1\!/_2O_2 + H_2O \rightarrow H_2SO_4 \quad (1.25)$$

NOx reduction with ammonia:

$$4NO + 4NH_3 + O_2 \rightarrow 4N_2 + 6H_2O \quad (1.26)$$

Char regeneration at 400 °C:

$$H_2SO_4 \rightarrow H_2O + SO_3 \quad (1.27)$$

$$2SO_3 + C \rightarrow 2SO_2 + CO_2 \quad (1.28)$$

After concentration, SO_2 is sent to the Claus unit. This process can achieve an SO_2 removal efficiency of 90 % and an NOx removal efficiency of 70 %.

1.7.5.5 Wellman–Lord Process

The Wellman–Lord process consists in the SO_2 neutralization by a sodium-based solution which is then regenerated. The following chemical reactions occur:

SO_2 capture:

$$SO_2 + Na_2SO_3 + H_2O \rightarrow 2NaHSO_3 \tag{1.29}$$

$$Na_2SO_3 + {}^1\!/_2O_2 \rightarrow Na_2SO_4 \tag{1.30}$$

Regeneration:

$$2NaHSO_3 \rightarrow SO_2 + Na_2SO_3 + H_2O \tag{1.31}$$

SO_2-rich gas treatment:

$$2SO_2 + 6H_2 \rightarrow 2H_2S + 4H_2O \tag{1.32}$$

$$2H_2S + SO_2 \rightarrow 3S + 2H_2O \tag{1.33}$$

The final effluent is sent to the Claus unit. This process has been often used for utility and industrial boilers. It has the advantage to regenerate the scrubbing solution and to produce a saleable product. However, installation and maintenance costs are higher than lime, limestone, or dual-alkali systems.

1.7.6 Oxides of Nitrogen, NOx

Three oxides of nitrogen are listed on Table 1.19. Nitric oxide, NO, nitrous oxide, N_2O, and nitrogen dioxide, NO_2, are formed in internal combustion engines and by blasting. The proportion of nitrous oxide is likely to be small. Furthermore, nitric oxide converts rapidly to nitrogen dioxide in the presence of air and water vapor.

$$2NO + O_2 \rightarrow 2NO_2 \tag{1.34}$$

As nitrogen dioxide is the most toxic of these oxides of nitrogen, it is sensible to concentrate on the physiological effects of this gas. At the temperatures found in underground openings, it is probable that nitrogen dioxide will be mixed with a companion gas, nitrogen tetroxide, N_2O_4, having similar physiological effects. The brown fumes of nitrogen dioxide dissolve readily in water to form both nitrous (HNO_2) and nitric (HNO_3) acids.

$$2NO_2 + H_2O \rightarrow HNO_2 + HNO_3 \tag{1.35}$$

Table 1.26 The progressive symptoms of oxides of nitrogen, NOx

Concentration of nitrogen dioxide, ppm	Effects
40	May be detected by smell
60	Minor throat irritation
100	Coughing may commence
150	Severe discomfort, may cause pneumonia later
200	Likely to be fatal

These acids cause irritation and, at higher concentrations, corrosive effects on the eyes and respiratory system. The progressive symptoms are shown in Table 1.26.

The immediate treatment for nitrogen dioxide poisoning is similar to that for sulfur dioxide, namely the administration of oxygen, immobility, and warmth. An insidious effect of nitrogen dioxide poisoning is that an apparent early recovery can be followed, soon afterward, by the development of acute bronchopneumonia.

1.7.6.1 Control Techniques for NOx Emissions Reduction

NOx reduction in boilers and process heaters can be achieved with combustion modification and flue gas treatment or a combination of these. The choice of the technique depends on the type and size of the boiler or heater, the fuel characteristics, and the flexibility for modifications. Practically, NOx reductions consist in *thermal NOx*[1] reduction and *fuel* NOx[2] reduction.

When fuel with low nitrogen content is used, such as fuel gas or distillate oil, thermal NOx is the only component that can be controlled.

Combustion control involves consequently three main strategies:

- Reducing peak temperatures in the combustion zone.
- Reducing the gas residence time in the high-temperature zone.
- Reducing oxygen concentrations in the combustion zone.

These changes can be achieved with process modifications or operating condition modifications. Finally, the flue gas treatment allows reducing NOx emissions.

Below, different technologies are generally and shortly described. The table synthesized information available concerning efficiency and applicability of these technologies on process heaters or boilers in petroleum industry, using fuel oil or fuel gas. Only methods that have been used for industrial process heaters or boilers are considered here, but many other techniques exist.

---

[1] Thermal NOx is produced by combination at high flame temperature of nitrogen and oxygen contained in the combustion air supply. It is produced during the combustion of both fuel gases and fuel oils.

[2] Fuel NOx is produced by combination of nitrogen contained in the fuel with excess oxygen contained in the combustion air. It is only a problem with fuel oils containing bound nitrogen.

1.7.6.2 Low-NOx Burners (LNB)

Low-NOx burner is a technology allowing a controlled mixing of fuel and air, resulting in a cooler flame and consequently less thermal NOx formation. It is designed as a stage combustion with either staged air or staged fuel. It is applicable to tangential and wall-fired boilers of various sizes and heaters. It reduces emissions from 40 to 60 %.

The basic principle of low-NOx burner is the separated injection of air and fuel in the furnace, resulting in the destruction of NOx in the flame (fuel-rich combustion zones) and the peak flame temperature suppression. Moreover, the better air flow distribution allows fuel ignition and flame stability.

1.7.6.3 Staged Combustion air (SCA)

Staged combustion air allows the reduction of fuel NOx by suppressing the amount of air below that required for complete combustion. It is achieved by injecting a portion of the total combustion air downstream of the fuel-rich primary combustion zone.

The SCA can be accomplished by several means such as burners out of service (BOOS), biased firing, or over fire air (OFA), depending on the type of boiler. The SCA technique is highly effective on high-nitrogen fuels such as residual oil. It reduces NOx emissions by 20–50 %.

1.7.6.4 Flue Gas Recirculation (FGR)

Flue gas recirculation consists in the rerouting of a portion of flue gases from the stack back to the furnace. Thus, furnace temperature and oxygen concentration are reduced and so is thermal NOx formation. Large modifications to the burner and wind box in old boilers are expensive so this technique is better for new boilers.

1.7.6.5 Water or Steam Injection (WI/SI)

Water or steam injection in the flame reduces thermal NOx formation by lowering the peak temperature of the flame. This technique has a relatively low initial cost so it is considered as quite efficient for smaller boilers. However, this technique can lead to thermal losses and increase in CO emissions.

1.7.6.6 Selective Non-Catalytic Reduction (SNCR)

SNCR is a post-combustion technique consisting in injecting ammonia or urea into combustion flue gases. A reaction with NOx occurs to produce nitrogen and water. There are not many experiences to evaluate effectiveness of this technique.

1.7.6.7 Selective Catalytic Reduction (SCR)

SCR is another post-combustion technique consisting in injecting ammonia into the combustion zone in the presence of a catalyst to reduce NOx into nitrogen and water. This method allows achieving NOx emission reduction by 75–90 %. This technique is rather common. Both SNCR and SCR are influenced by sulfur content of the flue gas.

1.7.7 Hydrogen Sulfide, H_2S

The presence of this highly toxic gas is readily detected by its characteristic smell of bad eggs. Unfortunately, hydrogen sulfide has a narcotic effect on the nervous system including paralysis of the olfactory nerves. Hence, after a short exposure, the sense of smell can no longer be relied upon.

Hydrogen sulfide is produced by acidic action or the effects of heating on sulfide ores. It is formed naturally by bacterial or chemical decomposition of organic compounds and may often be detected close to stagnant pools of water in underground mines. Hydrogen sulfide may occur in natural gas or petroleum reserves and migrate through the strata in a weakly acidic water solution. It can also be generated in gob fires. In such cases, free sulfur may be deposited by partial oxidation of the gas.

$$2H_2S + O_2 \rightarrow 2H_2O + 2S \tag{1.36}$$

This can sometimes be seen as a yellow deposit in burned areas. However, in a plentiful supply of air, hydrogen sulfide will burn with a bright blue flame to produce sulfur dioxide.

$$2H_2S + 3O_2 \rightarrow 2SO_2 + 2H_2O \tag{1.37}$$

The physiological effects of hydrogen sulfide may be listed as follows in Table 1.27.

A victim who recovers from hydrogen sulfide poisoning may be left with longer-term conjunctivitis and bronchitis.

Table 1.27 The physiological effects of hydrogen sulfide

Concentration of hydrogen sulfide, ppm	Effects
0.1–1	Detectable by smell
5	Beginning of toxicity
50–100	Slight irritation to eyes and respiratory tract, headache, loss of odor after 15 min
200	Intensified irritation of nose and throat
500	Serious inflammation of eyes, nasal secretions, coughing, palpitations, fainting
600	Chest pains due to corrosion of respiratory system, may be fatal
700	Depression, coma, probable death
1,000	Paralysis of respiratory system, very rapid death

1.7.8 Hydrogen, H_2

Although non-toxic, hydrogen is the most explosive of all the mine gases. It burns with a blue flame and has the wide flammable range of 4–74.2 % in air. Hydrogen can be ignited at a temperature as low as 580 °C and with ignition energy about half of that required by methane. Hydrogen occasionally appears as a strata gas and may be present in afterdamp at about the same concentrations as carbon monoxide.

Dangerous accumulations of hydrogen may occur at locations where battery charging is in progress. Hydrogen has a density only some 0.07 that of air. It will, therefore, tend to rise to the roof. Battery charging stations should be located in intake air with a duct or opening at roof level that connects into a return airway.

1.7.9 Radon, Rn

This chemically inert gas is one of the elements formed during radioactive disintegration of the uranium series. Although its presence is the most serious in uranium mines, it may be found in many other types of underground openings. Indeed, seepages of radon from the ground into the basements of surface buildings have been known to create a serious health hazard.

Radon emanates from the rock matrix or from groundwater that has passed over radioactive minerals. It has a half-life of 3.825 days and emits alpha radiation. The immediate products of the radioactive decay of radon are minute solid particles known as the radon daughters. These adhere to the surfaces of dust particles and emit alpha, beta, and some gamma radiation.

1.8 Gas Mixtures

Although the gases that exist most commonly in chemical and petroleum have been discussed separately in previous section, it is more usual that several gaseous pollutants appear together as gas mixtures. Furthermore, the importation of an ever-widening range of materials into facilities also introduces the risk of additional gases being emitted into the industrial environment. Special consideration should be given to the application of the TLVs in assessing the health hazards that may be associated with exposure to mixtures of two or more substances. A brief discussion of basic considerations involved in developing TLVs for mixtures and methods for their development is presented in the section below.

1.8.1 Threshold Limit Values for Gas Mixtures

TLVs refer to airborne concentrations of substances and represent conditions under which it is believed that nearly all workers may be repeatedly exposed day after day without adverse health effects.

These limits are intended for use in the practice of industrial hygiene as guidelines or recommendations in the control of potential health hazards and for no other use, e.g., in the evaluation or control of community air pollution nuisances; in estimating the toxic potential of continuous, uninterrupted exposures or other extended work periods; as proof or disproof of an existing disease or physical condition.

1.8.1.1 Definitions

Three categories of TLVs are specified herein, as follows:

1. Threshold limit value–time-weighted average (TLV-TWA)—the TWA concentration for a normal 8-h workday and a 40-h workweek, to which nearly all workers may be repeatedly exposed, day after day, without adverse effect.
2. Threshold limit value–short-term exposure limit (TLV-STEL)—the concentration to which workers can be exposed continuously for a short period of time without suffering from:
   (a) Irritation,
   (b) Chronic or irreversible tissue damage, or
   (c) Narcosis of sufficient degree to increase the likelihood of accidental injury, impair self-rescue, or materially reduce work efficiency, provided that the daily TLV-TWA is not exceeded. It is not a separate independent exposure limit; rather, it supplements the TWA limit where there are recognized acute effects from a substance whose toxic effects are primarily of a chronic nature. STELs are recommended only where toxic effects have been reported from high short-term exposures in either humans or animals.
   An STEL is defined as a 15-min TWA exposure which should not be exceeded at any time during a workday even if the 8-h TWA is within the TLV-TWA. Exposures above the TLV-TWA up to the STEL should not be longer than 15 min and should not occur more than four times per day. There should be at least 60 min between successive exposures in this range. An averaging period other than 15 min may be recommended when this is warranted by observed biological effects.
3. Threshold limit value–ceiling (TLV-C) the concentration that should not be exceeded during any part of the working exposure.

In conventional industrial hygienic practice if instantaneous monitoring is not feasible, then the TLV-C can be assessed by sampling over a 15-min period except for those substances that may cause immediate irritation when exposures are short.

For some substances, e.g., irritant gases, only one category, the TLV–ceiling, may be relevant. For other substances, one or two categories may be relevant, depending upon their physiological action. It is important to observe that if any one of these types of TLVs is exceeded, a potential hazard from that substance is presumed to exist.

TLVs based on physical irritation should be considered no less binding than those based on physical impairment. There is increasing evidence that physical irritation may initiate, promote, or accelerate physical impairment through interaction with other chemical or biological agents.

TWAs permit excursions above the TLV, provided they are compensated by equivalent excursions below the TLV-TWA during the workday. In some instances, it may be permissible to calculate the average concentration for a workweek rather than for a workday.

The relationship between the TLV and permissible excursion is a rule of thumb and in certain cases may not apply. The amount by which the TLVs may be exceeded for short periods without injury to health depends upon a number of factors such as the nature of the contaminant, whether very high concentrations—even for short periods—produce acute poisoning, whether the effects are cumulative, the frequency with which high concentrations occur, and the duration of such periods. All factors must be taken into consideration in arriving at a decision as to whether a hazardous condition exists.

In order to widen the applicability of this section, the recommended TLVs of a number of other gases and vapors are given in Table 1.28 and grouped in terms of probable sources. Again, we are reminded that these TLVs are simply guidelines and that national or state mandated limits should be consulted to ensure compliance with the relevant laws.

In any given atmosphere, if there are two or more airborne pollutants (gaseous or particulate) that have adverse effects on the same part of the body, then the TLV should be assessed on the basis of their combined effect. This is calculated as the dimensionless sum.

$$\frac{C_1}{T_1} + \frac{C_2}{T_2} + \ldots\ldots \frac{C_n}{T_n} \tag{1.38}$$

where C = measured concentration

and T = corresponding threshold limit value.

If the sum of the series exceeds unity, then the TLV of the mixture is deemed to be exceeded.

1.8.1.2 Example:

An analysis of air samples taken in a return airway indicates the following gas concentrations:

carbon dioxide: 0.2 %,
carbon monoxide: 10 ppm,
hydrogen sulfide: 2 ppm,
sulfur dioxide: 1 ppm.

Determine the TLVs, TWA and STEL for the mixture.

Sulfur dioxide and hydrogen sulfide are both irritants to the eyes and respiratory system. Furthermore, carbon dioxide is a stimulant to breathing, increasing the rate of ventilation of the lungs. Hence, it may be regarded as being synergistic with the sulfur dioxide and hydrogen sulfide.

Carbon monoxide, however, affects the oxygen-carrying capacity of the bloodstream and need not be combined with the other gases in determining the

Table 1.28 Threshold limit values for other gases and vapors that may be present underground

Substance	Guideline time-weighted average limit (ppm unless otherwise stated)
Cleaners and solvents	
Acetone	750 (STEL = 1,000)
Ammonia	Ammonia 25 (STEL = 35)
Toluene	Toluene 100 (STEL = 150)
Turpentine	Turpentine 100 (STEL = 150)
Refrigerants	
Ammonia	25 (STEL = 35)
R11	R11 1000 (ceiling)
R12	R12 1000
R22	R22 1000
R112	R112 500
Fuels	
Butane	800
Gasoline vapor	Gasoline vapor 300 (STEL = 500)
Liquid petroleum gas	Liquid petroleum gas 1,000
Naphtha (coal tar)	Naphtha (coal tar) 100
Pentane	Pentane 600 (STEL = 750)
Propane	Propane 1,000
Welding and soldering	
General welding fumes	5 mg/m^3
Iron oxide fumes	Iron oxide fumes 5 mg/m^3
Lead fumes	Lead fumes 0.15 mg/m^3
Ozone (arc welding)	Ozone (arc welding) 0.1 ppm ceiling
Fluorides (fluxes)	Fluorides (fluxes) 2.5 mg/m^3
Heated plastics	
Carbon dioxide	Carbon dioxide 0.5 % (STEL = 3.0 %)
Carbon monoxide	Carbon monoxide 50 (STEL = 400)
Hydrogen chloride fumes	Hydrogen chloride fumes 5 ppm ceiling
Hydrogen cyanide	Hydrogen cyanide 10
Hydrogen fluoride	Hydrogen fluoride 3 ppm ceiling
Phenol (absorbed by skin)	Phenol (absorbed by skin) 5

(continued)

Table 1.28 (continued)

Substance	Guideline time-weighted average limit (ppm unless otherwise stated)
Explosives	
Carbon dioxide	0.5 % (STEL = 3.0 %)
Carbon monoxide	50 (STEL = 400)
Oxides of nitrogen	25
Ammonia	25 (STEL = 35)
Sulfur dioxide	2 (STEL = 5)
Nitrous acid fumes	
Nitric acid fumes	2 (STEL = 4)
Others	
Chlorine (biocides)	0.5 (STEL = 1)
Cresol (wood preservative)	5
Mercury vapor	0.05 mg/m^3
Oil mist (mineral): vapor free	5 mg/m^3
Oil mist (vegetable): vapor free	10 mg/m^3
Sulfuric acid fumes (batteries)	1 mg/m^3

TLVs for the mixture. From Table 1.23, the following TWAs and STELs are determined in Table 1.29.

The carbon monoxide concentration measured at 10 ppm and treated separately is shown to be less than the TWA of 50 ppm. The equivalent TLV of the remainder of the mixture is assessed from equation:

$$\text{TWA:} \ \frac{0.2}{0.5} + \frac{2}{10} + \frac{1}{2} = 1.1 \tag{1.39}$$

Table 1.29 Calculated time-weighted averages (TWA) and short-term exposure limits (STEL)

Component	Threshold limit values, TWA	Threshold limit values, STEL
Carbon dioxide	0.5 %	3 %
Hydrogen sulfide	10 ppm	15 ppm
Sulfur dioxide	2 ppm	5 ppm
Carbon monoxide	50 ppm	400 ppm

$$\text{STEL: } \frac{0.2}{3} + \frac{2}{15} + \frac{1}{5} = 0.4 \tag{1.40}$$

The dimensionless TWA is greater than 1. Hence, the TWA limit is exceeded. However, the STEL is less than unity. The implication is that personnel may spend time periods not exceeding 15 min in this atmosphere, but not a complete 8-h shift.

In some cases of gas mixtures, it may be practicable to monitor quantitatively for one pollutant only, even though it is known that other gases or particulates are present. In such circumstances, a pragmatic approach is to reduce the TLV of the measured substance by a factor that is assessed from the number, toxicity, and estimated concentrations of other contaminants known to be present.

1.8.2 Excursion Limits

For the vast majority of substances with a TLV-TWA, there is not enough toxicological data available to warrant a short-term exposure limit (STEL). Nevertheless, excursions above the TLV-TWA should be controlled even where the 8-h TLV-TWA is within recommended limits. Earlier editions of the TLV list included such limits whose values depended on the TLV-TWAs of the substance in question.

While no rigorous rationale was provided for these particular values, the basic concept was intuitive.

In a well-controlled process exposure, excursions should be held within some reasonable limits. Unfortunately, neither toxicology nor collective industrial hygiene experience provides a solid basis for quantifying what those limits should be. The approach here is that the maximum recommended excursion should be related to variability generally observed in actual industrial processes. In reviewing large numbers of industrial hygiene surveys, it is found that short-term exposure measurements were generally log normally distributed with geometric standard deviations mostly in the range of 1.5–2.0.

While a complete discussion of the theory and properties of the lognormal distribution is beyond the scope of this section, a brief description of some important terms is presented.

The measure of central tendency in a lognormal description is the antilog of the mean logarithm of the sample values. The distribution is skewed, and the geometric mean is always smaller than the arithmetic mean by an amount, which depends on the geometric standard deviation. In the lognormal distribution, the geometric standard deviation is the antilog of the standard deviation of the sample value logarithms.

If a process displays a variability greater than this, it is not under good control and efforts should be made to restore control. This concept is the basis for the following excursion limit recommendation which apply to those TLV-TWAs that do not have STELS.

Excursions in worker exposure levels may exceed three times the TLV-TWA for no more than a total of 30 min during a workday, and under no circumstances should they exceed five times the TLV-TWA, provided that the TLV-TWA is not exceeded.

The approach is a considerable simplification of the idea of the lognormal concentration distribution but is considered more convenient to use by the practicing industrial hygienist. If exposure excursions are maintained within the recommended limits, the geometric standard deviation of the concentration measurements will be near 2.0 and the goal of the recommendations will be accomplished.

When the toxicological data for a specific substance are available to establish an STEL, this value takes precedence over the excursion limit regardless of whether it is more or less stringent.

1.8.3 "Skin" Notation

The designation "Skin" in the "Notations" column refers to the potential significant contribution to the overall exposure by the cutaneous route, including mucous membranes and the eyes, either by contact with vapors or, of probable greater significance, by direct skin contact with the substance. Vehicles present in solutions or mixtures can also significantly enhance potential skin absorption.

It should be noted that while some materials are capable of causing irritation, dermatitis, and sensitization in workers, these properties are not considered relevant when assigning a skin notation. It should be noted, however, that the development of a dermatological condition can significantly affect the potential for dermal absorption.

While relatively limited quantitative data currently exist with regard to skin absorption of gases, vapors, and liquids by workers, it is recommended that the integration of data from acute dermal studies and repeated dose dermal studies in animals and/or humans, along with the ability of the chemical to be absorbed, be used in deciding on the appropriateness of the skin notation.

In general, available data which suggest that the potential for absorption via the hands/forearms during the workday could be significant, especially for chemicals with lower TLVs, could justify a skin notation. From acute animal toxicity data, materials having a relatively low dermal LD50 (1,000 mg/kg of body weight or less) would be given a skin notation. Where repeated dermal application studies have shown significant systemic effects following treatment, a skin notation would be considered.

When chemicals penetrate the skin easily (higher octanol–water partition coefficients) and where extrapolations of systemic effects from other routes of exposure suggest that dermal absorption may be important in the expressed toxicity, a skin notation would be considered.

Substances having a skin notation and a low TLV may present special problems for operations involving high airborne concentrations of the material, particularly under conditions where significant areas of the skin are exposed for a long period of time. Under these conditions, special precautions to significantly reduce or preclude skin contact may be required.

Biological monitoring should be considered to determine the relative contribution of exposure via the dermal route to the total dose.

1.8.3.1 "Sensitizer" Notation

The designation "SEN" in the "Notations" column refers to the potential for an agent to produce sensitization, as confirmed by human or animal data. The SEN notation does not imply that sensitization is the critical effect on which the TLV is based, nor does it imply that this effect is the sole basis for that agent's TLV. If sensitization data exist, they are carefully considered when recommending the TLV for the agent. For those TLVs that are based upon sensitization, they are meant to protect workers from induction of this effect and are not intended to protect those workers who have already become sensitized.

In the workplace, respiratory, dermal, or conjunctival exposures to sensitizing agents may occur. Similarly, sensitizers may evoke respiratory, dermal, or conjunctival reactions. At this time, the notation does not distinguish between sensitization involving any or these organ systems. The absence of an SEN notation does not signify that the agent lacks the ability to produce sensitization but may reflect the paucity or inconclusiveness of scientific evidence.

Sensitization often occurs via an immunological mechanism and is not to be confused with other conditions or terminology such as hyperreactivity, susceptibility, or sensitivity. Initially, there may be little or no response to a sensitizing agent. However, after a person is sensitized, subsequent exposure may cause intense responses even at low exposure concentrations (well below the TLV).These reactions may belief-threatening and may have an immediate or delayed onset. Workers who have become sensitized to a particular agent may also exhibit cross-reactivity to other agents with similar chemical structures.

A reduction in exposure to the sensitizer and its structural analogs generally reduces the incidence of allergic reactions among sensitized individuals. For some sensitized individuals, however, complete avoidance in occupational and non-occupational settings provides the only means to prevent the immune responses to recognized sensitizing agents and their structural analogs.

Agents having an SEN notation and a low TLV present special problems in the workplace. Respiratory, dermal, and conjunctival exposures should be significantly reduced or eliminated using personal protective equipment and process control measures. Education and training (e.g., review of potential health effects, safe handling procedures, emergency information) are also necessary for those who work with known sensitizing agents.

For additional information regarding the sensitization potential of a particular agent, refer to the Documentation for the specific agent.

Although the TWA concentration provides the most satisfactory, practical way of monitoring airborne agents for compliance with the TLVs, there are certain substances for which it is inappropriate. In the latter group are substances which are predominantly fast acting and whose TLV is more appropriately based on this particular response. Substances with this type of response are best controlled by a ceiling limit that should not be exceeded.

It is implicit in these definitions that the manner of sampling to determine non-compliance with the limits for each group must differ; a single, brief sample, which is applicable to a ceiling limit, is not appropriate to the TWA; here, a sufficient number of samples are needed to permit a TWA concentration throughout a complete cycle of operations or throughout the work shift.

Whereas the ceiling limit places a definite boundary that concentrations should not be permitted to exceed, the TWA requires an explicit limit to the excursions that are permissible above the listed TLVs. It should be noted that the same factors are used by the Chemical Substances TLV Committee in determining the magnitude of the value of the STELs, or whether to include or exclude a substance for a ceiling listing.

1.8.4 Simple Asphyxiants -"Inert" Gases of Vapors

A number of gases and vapors, when present in high concentrations in air, act primarily as simple asphyxiants without other significant physiological effects. A TLV may not be recommended for each simple asphyxiant because the limiting factor is the available oxygen. The minimal oxygen content should be 18 % by volume under normal atmospheric pressure (equivalent to a partial pressure, pO_2 of 135 torr). Atmospheres deficient in O_2 do not provide adequate warning, and most simple asphyxiants are odorless. Several simple asphyxiants present an explosion hazard. Account should be taken of this factor in limiting the concentration of the asphyxiant.

1.8.5 Biological Exposure Indices (BEIs)

The note "BEI" is listed in the "Notations" column when a BEI is also recommended for the substance listed. Biological monitoring should be instituted for such substances to evaluate the total exposure from all sources, including dermal, ingestion, or non-occupational. See the BEI section in this book and the Documentation of the TLVs and BEIs for the substance.

1.8.6 Physical Factors

It is recognized that such physical factors as heat, ultraviolet and ionizing radiation, humidity, abnormal pressure (altitude), and the like any place added stress on the body so that the effects from exposure at a TLV may be altered. Most of these stresses act adversely to increase the toxic response of a substance. Although most TLVs have built-in safety factors to guard against adverse effects to moderate deviations from normal environments, the safety factors of most substances are not of such a magnitude as to take care of gross deviations. For example, continuous, heavy work at temperatures above 25 °C or overtime extending the workweek more than 25 % might be considered gross deviations. In such instances, judgment must be exercised in the proper adjustments of the TLVs.

1.8.7 Unlisted Substances

The list of TLVs is by no means a complete list of all hazardous substances used in industry. For a large number of materials of recognized toxicity, little or no data are available that could be used to establish a TLV. Substances that do not appear on the TLV list should not be considered to be harmless or non-toxic. When unlisted substances are introduced into workplace, the medical and scientific literature should be reviewed to identify potentially dangerous toxic effects. It may also be advisable to conduct preliminary toxicity studies. In any case, it is necessary to remain alert to adverse health effects in workers which may be associated with the use of new materials.

1.8.8 Particulates Not Otherwise Classified

In contrast to fibrogenic dusts which cause scar tissue to be formed in lungs when inhaled in excessive amounts, so-called nuisance dusts have a long history of little adverse effect on lungs and do not produce significant organic disease or toxic effect when exposures are kept under reasonable control. Such dusts have also been called (biologically) "inert" dusts, but the latter term is inappropriate to the extent that there is no dust which does not evoke some cellular response in the lung when inhaled in sufficient amount.

However, the lung–tissue reaction caused by inhalation of particulates not otherwise classified (PNOCs) has the following characteristics:

1. The architecture of the air spaces remains intact;
2. Collagen (scar tissue) is not formed to a significant extent; and
3. The tissue reaction is potentially reversible.

Excessive concentrations of PNOCs in the workroom air may seriously reduce visibility; may cause unpleasant deposits in the eyes, ears, and nasal passages (e.g., Portland cement dust); or may cause injury to the skin or mucous membranes by chemical or mechanical action or by the rigorous skin cleansing procedures necessary for their removal.

A TLV-TWA of 10 mg/m^3 of total dust containing no asbestos and <1 % crystalline silica is recommended for substances in these categories and for which no specific TLVs have been assigned.

1.9 Measurements of General Pollution of the Air

The effects of air pollution on human health, on plants, and on animals are well known. To ensure that all important methods of emission reduction have been considered, this section is prepared to provide information which will be of values to all concerned with measurements of general pollution of the air and involved in reducing to a realistic minimum the emission from industrial processes. It is an advantage that there are accepted methods for the measurement so that a sound comparison of all results could be made.

Next section specifies the minimum requirements for material, operation, tests, inspections and cover site and process measurements, calibration and installation of instruments.

1.10 Site and Process Measurement of Air Pollution

Site and process measurement is divided in two parts as follows:
Part 1: Fixed-Type Measurement
Part 2: Portable-Type Measurement

1.10.1 Fixed-Type Measurement

1.10.1.1 Fixed-type or Online Monitoring

One of the fastest and easiest process analyzer for petrochemical, refinery, and environmental monitoring is mass spectrometer. In general, this kind of instrument

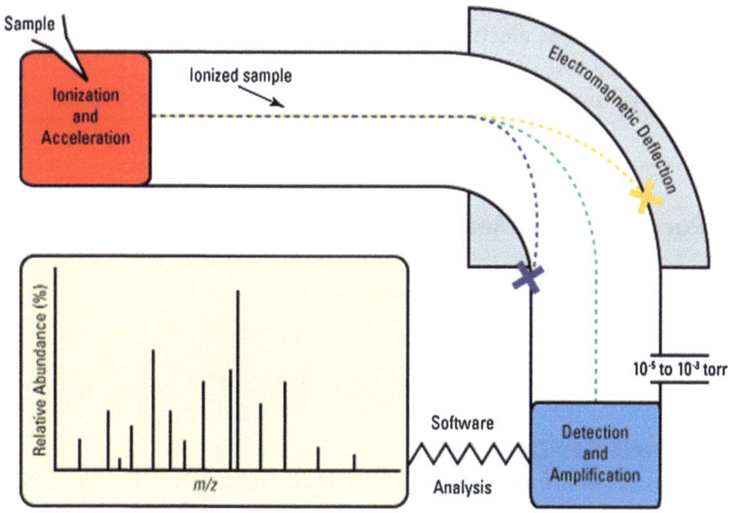

Fig. 1.13 Schematic diagram of an analytical mass spectrometer

is designed for measurement of ionizable gas. Sampling system is bleed solenoid valve.

In mass spectrometry, the sample should be converted into positive ions which are then separated and characterized. Figure 1.13 shows schematically the essential parts of a typical analytical mass spectrometer.

Construction and installation of instrument should be according to manufacturer's advice and purchase requirement, but in general, it should be in safe place with all necessary items such as electrical power, gas, and water supply. Also, it should be away from contaminated gas, shock, vibration, ignition of flammable gas.

Operation of instrument after test performance should be according to manufacturer's manual operation booklet.

All requirements should be according to supplier's advice such as electrical power carrier gas, standard gas, water supply, safe place, drainage, test equipment, and test run.

Calibration of instrument should be according to operation manual book and this could be monthly or weekly. All items for the calibration should be supplied by manufacturers.

Another fixed-type apparatus which is normally used in industry is SO_2, NO_x, and O_2 measuring analyzer. The principle of this instrument is based on split-beam photometric analysis. In this instrument, the process diode array is used as detector for monitoring SO_2 and NO_x. It could provide direct measurement of NO and NO_2 simultaneously.

Construction and installation of instrument should be according to supplier's recommendations. It could be installed even 305 m (1,000 feet) away for continuous emission monitoring. All requirements should be according to manufacturer's advice.

Operation of instrument after test performance should be according to manufacturer's manual operation booklet. All requirements should be according to supplier's advice such as electrical power, carrier gas, standard gas, water supply, test equipment, and test run.

Calibration of instrument should be based on supplier's manual operation booklet.

1.10.2 Portable-Type Measurement

The apparatus are used for measurement of

(a) Toxic gas detection
(b) Hydrocarbon gas detection
(c) CO_2, CO, and O_2 detection

1.10.2.1 Toxic Gas Detection

This apparatus is used for the detection of H_2S, HCN, Cl_2, CO_2, SO_2, NH_3, HF, HCl, NO_2, H_2, CO, F_2, Br_2, AsH_3, PH_3, SiH_4, B_2H_6, and GeH_4.

The basic principle of system is electrochemical diffusion sensors. There is no need for installation. It could be operated from −40 to +40 °C (−40 to 104 °F). There are different types of sensors for each gas to be measured and used. Operation of instrument after test performance should be according to supplier's manual operation booklet.

In general, this type of instrument is factory calibrated, and in case of failure, it should be returned to the manufacturer, or alternatively, only the user's competent personnel of instrument control may calibrate the instrument as per the instructions given in the supplier's manual operation.

1.10.3 Hydrocarbon Gas Detection

This system is provided for the detection of hydrocarbon gases such as natural gas ethylene, propane, and butane, as well as low concentration of chlorinated compounds.

Detection of system is flame ionization detector (FID), (see Fig. 1.14) which has been used widely in gas chromatography.

It is the premier detector in gas chromatography. It has unique properties and performance that puts it above and beyond all other general-use detectors in gas chromatography (or any other form of chromatography, for that matter). There is no need for installation, but for each gas, different devices have been used.

Operation of instrument after test performance should be in accordance with manufacturer's manual operation booklet. As for performance, the FID will locate gas leaks down to 1 ppm for early warnings of potentially dangerous conditions.

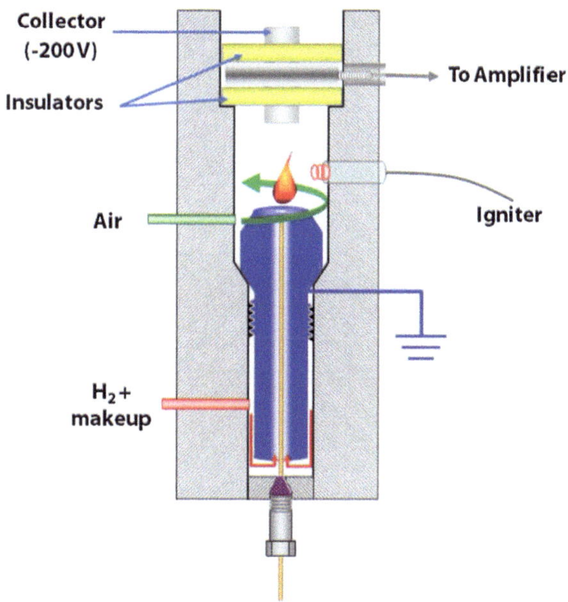

Fig. 1.14 Flame ionization detector (FID) for the detection of hydrocarbon gases

In general, this kind of instrument is factory calibrated. In case of failure, the instrument should be returned to the manufacturer or may be calibrated by competent personnel of user's instrument control in accordance with the guidelines given by the manufacturer.

1.10.4 CO_2, CO and O_2 Measurements

The Orsat apparatus is used for easy measurement of CO_2, CO, and O_2 gases. There are three bottles with glasses and one buret. Bottle P′ is filled with potassium or sodium hydroxide solution. Bottles P′, P″ are filled with potassium pyrogallate and cuprous chloride solutions, respectively. Inquiries should be made from manufactures in respect of material specification before ordering.

The basic structure of instrument is shown in Fig. 1.15.

Operation of this apparatus should be performed according to the supplier's manual operation booklet, but all solutions should be prepared in time of use.

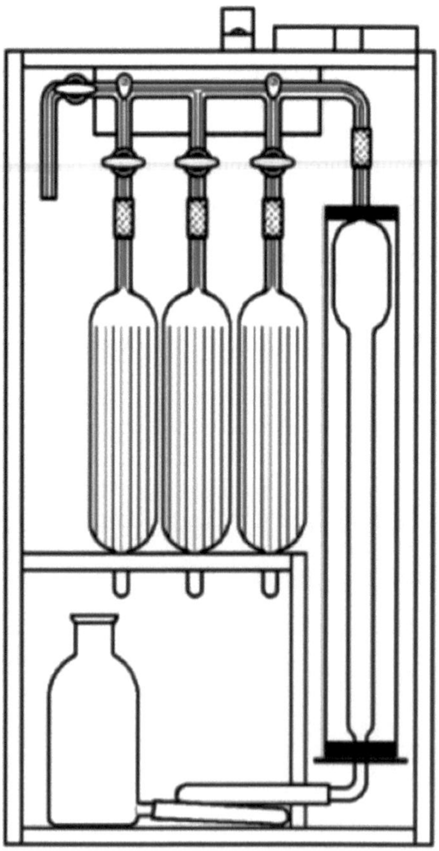

Fig. 1.15 Basic structure of Orsat apparatus

Instrument should be calibrated with a mixture of known gases. The instructions given by the manufacturer in the calibration of instrument shall be considered.

1.11 Sampling and Calibration of Gaseous Pollutants

One of the easiest ways of sampling which should be used is grab sampling. Grab sampling is a technique commonly used in air pollution investigation. In this type of test, a volume of air that can later be analyzed is required.

The containers are usually made from flexible plastics, steel, glass, or hypodermic syringes. Further specifications should be obtained from manufacturers before ordering. Care shall be taken in the selection of the syringe material to prevent contamination of gaseous samples. Repeated preconditioning treatments are necessary to avoid irreversible adsorption of the pollutant. Before using the syringe, the performance should be inspected to make sure it is in working order.

The basic principal structure of grab sampling is shown in Fig. 1.16.

1.12 Adsorption Sampling

Adsorbents are used to isolate a variety of pollutant gases. The materials, which should be used as adsorbent, are activated aluminas, silica gel, molecular sieve, charcoal, and celite. Further specifications should be obtained from manufacturers for purchaser consideration.

The sample should be drawn through a container with the adsorbent maintained at ambient or subambient temperatures. Adsorbed material can be removed by heating or washing with suitable solvent.

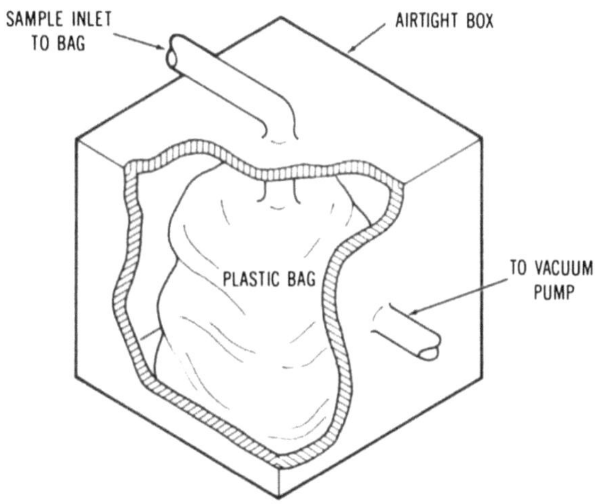

Fig. 1.16 A plastic bag being used for grab sampling

If water vapor is present in the sample stream, the adsorbent can be deactivated. Because of the nature of certain gases and adsorbents, it is not always possible to remove the adsorbed gas without decomposition. In addition, certain adsorbents are known to cause isomerization.

Absorption is the process of transferring one or more gaseous components into a liquid or solid medium in which they dissolve.

1.13 Bubblers and Impingers

This is a kind of equipment to trap specific atmospheric gases in a solution which should later be analyzed. The gas to be analyzed is drawn through a tube, the downstream end of which should be blown to the surface of a liquid. The dispersion tube should be an open-ended tube or one with a frit (material with many 50–100 μm holes).

Frits should be glass, plastic, and ceramic. For further material specifications, inquiries should be made from manufacturers before ordering.

A number of designs for bubblers and impingers are shown in Fig. 1.17.

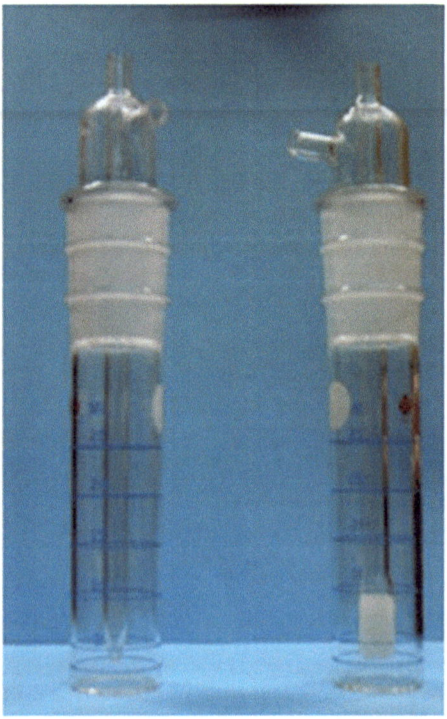

Fig. 1.17 Typical bubbler and impinger sampling equipment

1.14 Cryogenic Sampling

The cryosampling technique is designed to preconcentrate chemical or called cold trapping. Coolants frequently used for the trap are ice water, solid carbon dioxide, isopropyl alcohol, liquid oxygen, liquid nitrogen, and liquid helium.

Other equipments for this kind of sampling are pumps, sequential samplers, and operating power. Material specification should be obtained from manufacturers for purchaser consideration.

The cryogenic air sampling method is a way by use of the principle of cryogenic pumping. A cryogenic pump, which uses a cryogenic coolant, can continue to evacuate because most major components in the atmosphere are condensed to solid immediately. Liquid helium or liquid neon is candidates for the coolant in many cases. A sampling system using the cryogenic pump does not require any mechanical pumping system nor power; it is suitable for an onboard sampler. The sample amount is, generally speaking, not limited by the sampling altitude. When the volume of the sample cylinder is the same as that of the grab sampling, the sample amount of 1,000 times can be achieved easily by this method even over 30 km altitude.

As liquid neon has larger latent heat than liquid helium, it can condense air more efficiently. Neon, hydrogen, and helium exist as gases at the boiling point of liquid neon. Those gases are rare in the atmosphere; therefore, they have no effect on concentration measurements. Liquid helium temperature is, on the other hand, less than that of liquid neon, and only hydrogen and helium exist as gases; the saturation vapor pressure of hydrogen at the boiling point of liquid helium is merely 0.0001 Pa.

Though liquid helium has less latent heat than liquid neon and is required much in amount to sample the same amount of air, its heat capacity, mainly by sensible heat, can be utilized when heat exchange in the low-temperature gas is made efficient. Helium in the atmosphere cannot be condensed, but it has no effect on the concentration measurement of minor constituents because its amount decreases rapidly in proportion to the atmospheric pressure.

Helium can be sampled and stored when other atmospheric components are inhaled into the cylinder and when it is prevented from back diffusion.

While the coolant exists in the cryogenic sampling system, ambient air can be inhaled into the sample cylinder. As a result, much larger amount of air can be recovered than that by the grab sampling explained above. Moreover, the sample cylinder may be smaller and the possibility of adsorption and/or reaction on the inner surface be reduced dramatically.

Should the sample amount be not controlled appropriately, the air in the cylinder may become very high to compromise the safety when the air warms up to room temperature. There are technical problems awaiting solution in addition to those for the grab sampling, including the high pressure and the cryogenic technologies. General precautions for minor constituents' research are the same as

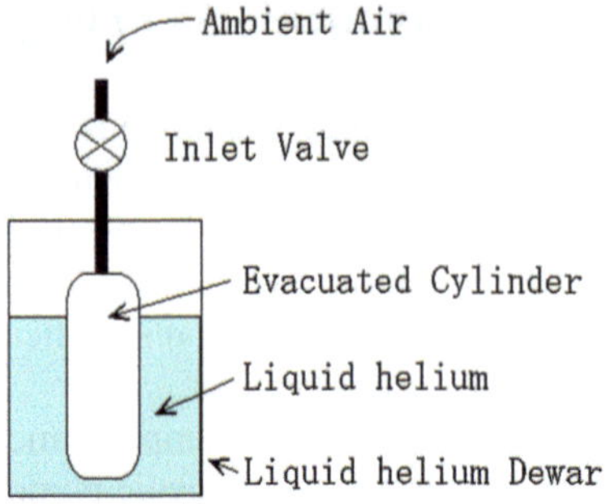

Fig. 1.18 The basic structure for cryosampling

those of the grab sampling. Table 1.26 shows the details of various coolants which can be used in cold bath solutions.

The basic structure for cryosampling is shown in Fig. 1.18.

1.15 Hydrocarbon and Carbon Dioxide

Non-dispersive infrared (NDIR) detectors are the industry standard method for measuring the concentration of carbon oxides (CO & CO_2).

Each constituent gas in a sample will absorb some infrared at a particular frequency. By shining an infrared beam through a sample cell (containing CO or CO_2), and measuring the amount of infrared absorbed by the sample at the necessary wavelength, an NDIR detector is able to measure the volumetric concentration of CO or CO_2 in the sample.

A chopper wheel mounted in front of the detector continually corrects the offset and gain of the analyzer and allows a single sampling head to measure the concentrations of two different gases. The combustion fast NDIR uses a unique sampling system, coupled to miniaturized NDIR technology to give millisecond response times. The combustion fast NDIR has two remote sampling heads controlled by a main control unit and is capable of sampling CO & CO_2 simultaneously in two locations.

Carbon oxides are air pollutant. For measurement of these gases in laboratory, new instrumental technique can be used; in addition, other methods in ASTM could be used.

NDIR analyzer is based on the absorption of infrared energy by the contaminant gas. This instrument consists of a sample cell and reference cell, two infrared sources, and detector cell. Reference cell is sealed and contains a gas transparent to infrared wavelengths. Further specification should be obtained from manufacturers for purchaser consideration.

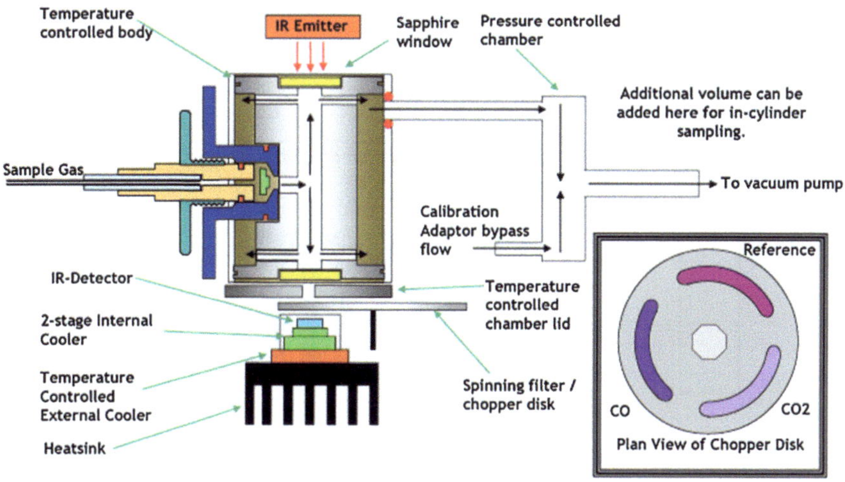

Fig. 1.19 Non-dispersive infrared (NDIR) analyzer

The basic structure of instrument is shown in Fig. 1.19. Water vapor is interference in carbon monoxide analysis; it is necessary to dry the sample by refrigeration or the use of drying agents.

1.16 Gas Chromatography

Gas chromatography method is used for measurement of total hydrocarbons, methane, and carbon monoxide (CO).

Carrier gas which is helium carries gas through column. Column, which is packed with different material, causes separation of gases.

These gases after separation reach to detector and with using FID; the different components could be analyzed. Various carrier gases and detectors could be used which depends on the method of analysis.

Further material specifications should be obtained from manufacturers for purchaser consideration.

A block diagram of a gas chromatography system is shown in Fig. 1.20.

Construction and installation of instrument should be according to supplier's recommendations. Operation of instrument after test performance should be according to supplier's manual operation booklet.

All requirements should be according to supplier's advice, such as electrical power, carrier gas, standard gas, water supply, test equipment, and test run.

Calibration of instrument should be according to supplier's manual operation booklet.

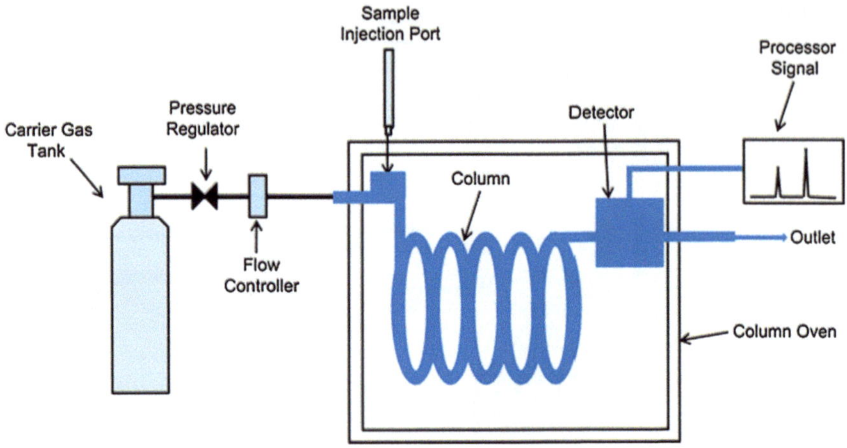

Fig. 1.20 Gas chromatographic analyzer for total hydrocarbons (THC), methane (CH_4), and carbon monoxide (CO)

1.17 Sulfur Compounds

Sulfur compounds are air pollutants. For measurement of these gases in laboratory, several methods and techniques have been used, which could be divided in the following parts.

There is no monitoring system for sulfur oxides which is used for stationary emission sources, but automated analyzers for sulfur dioxide are commercially available for this purpose. They employ various measuring principles including the NDIR absorptiometry, the ultraviolet absorptiometry, the gas dissolution conduct metric method, the ultraviolet fluorescence method, the controlled potential electrolysis method, and the interferometry.

At present, most automated analyzers for sulfur dioxide use NDIR absorptiometry. Sulfur dioxide is easily dissolved in condensation water and well adsorbed. Therefore, special consideration must be given to the configuration of the pretreatment unit, the materials exposed to gases, and the other factors.

The gas dissolution conduct metric method is used to determine the concentration of sulfur dioxide by allowing a sulfuric solution containing hydrogen peroxide to absorb sulfur dioxide in the sample gas (so that it becomes sulfuric acid) and then measuring the increase in conductivity due to this sulfuric acid.

The absorption solution is intermittently sent to the absorption pipe using a pump, and then, all owed to spontaneously drip into the absorption pipe. In the process of going to the bottom of the reaction pipe, the drops absorb a certain quantity of the sample gas. Sulfuric dioxide is absorbed by and reacts with the drops.

The resulting solution is sent to the measurement cell, and its conductivity is measured to obtain the concentration of sulfur dioxide. This method is applicable if the interference effect of any other alkaline or acid substance can be neglected or eliminated. The response speed is slow.

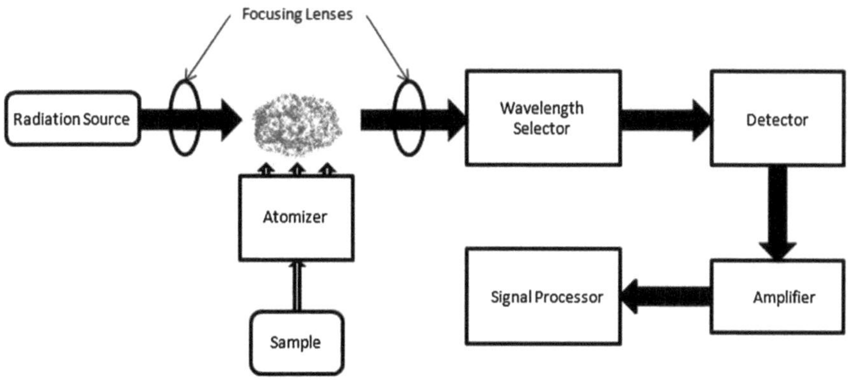

Fig. 1.21 Schematic diagram of atomic absorption spectrometer

1.18 Sampling of Particulate Matter

Airborne particulates or aerosols consist of either liquid or solid particles ranging in diameter from 0.01 μm or less, up to 100 μm. These should be isolated by sedimentation, filtration, impingement, centrifugation, or electrostatic or thermal precipitation. For more detail, refer to ASTM Vol. 11.03, Section 11, 1989.

After sampling of particulate matter for the measurement of inorganic element, there are several methods to be used. In this section, two important methods are described. For other methods, reference is made to ASTM Vol. 11.03, Section 11, 1989.

In an atomic absorption analysis, the element being determined must be reduced to the elemental state, vaporized, and imposed in the beam of radiation from the source.

An instrument has components as a spectrophotometer. These include a source, a monochromator, a sample container, a flame, a detector, and an amplifier indicator. Detail of specification can be obtained from manufacturers for purchaser consideration.

The basic element of instrument is shown in Fig. 1.21.

1.19 Inductively Coupled Argon Plasma Emission Spectroscopy

Gaseous ions or molecules, when thermally or electrically excited, emit characteristic radiation in the ultraviolet and visible regions. Emission spectroscopy is concerned with the characterization of the wavelengths and the intensities of radiation produced in this manner.

An instrument has components as a spectrophotometer. These include a plasma source, a monochromator, a sample container, a detector, and an amplifier indicator.

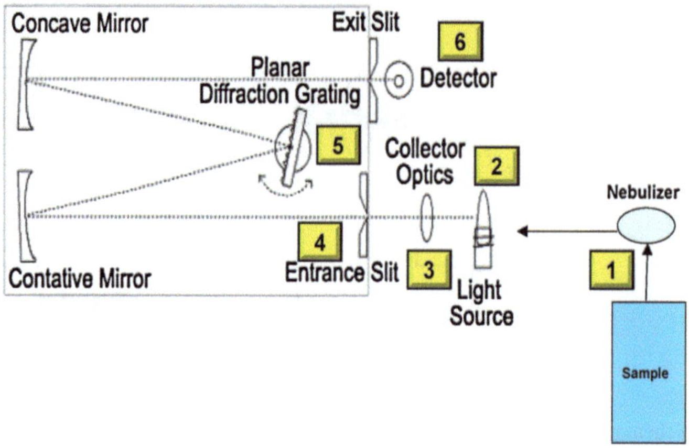

Fig. 1.22 ICP optical schematic

The basic structure of instrument is shown in Fig. 1.22.

1.20 Particulate Removal

- **Selection of Equipment for Particulate Removal**

Particulate matter (dust and grit) comes in great varieties of size, grain loading, shape, chemical composition, specific gravity, etc. In general, there are four basic types of equipment available for particulate removal:

(a) mechanical collectors;
(b) fabric filters;
(c) wet scrubbers; and
(d) electric precipitators (see also ISO 6584).

1.20.1 Mechanical Collector

1.20.1.1 Gravity Settling Chamber

This collector slows the gas from conveying velocities to settling velocities. Dust could be settled under influence of gravity.

Material for this kind of collector includes pneumatic pumps, nozzles, hoppers, electrical power. Specification can be obtained from manufacturers.

1.20.2 Construction

A typical gravity settling chamber is shown in Fig. 1.23. This type of mechanical collector has very low collection efficiencies on fine and moderately fine dusts.

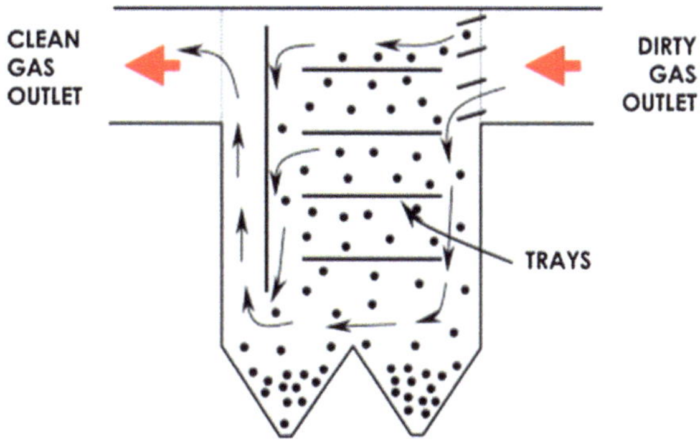

Fig. 1.23 Equipment for particulate removal (gravity settling chamber)

Installation and inspection should be as specified by manufacturers in their manual books.

1.21 Recirculating Baffle Collector

This type of collector where the gas is to be cleaned is introduced at high velocity under a horizontal baffle. Materials for this kind of collector include pneumatic pumps, pipe cleaning baffle, recirculating flow control baffle, dust slot, hoppers, and electrical power.

A typical recirculating baffle collector is shown in Fig. 1.24. In this type of collector, gas is introduced at high velocity under a horizontal baffle made up of rods spaced at 12.5 mm (half-inch) apart. The circulating flow is controlled at a nominal velocity by the expanding dust slot and the circulating flow control baffle.

Installation and inspection should be as specified by the manufacturers in their manual books.

1.22 High-Efficiency Cyclones

High-efficiency cyclones are inertial separators which employ a rotating flow. The particles are separated under the effect of centrifugal forces from the gaseous flow and are released in a radial direction.

(a) **Material**

Cyclone collector can be manufactured from a wide range of materials including mild steel, low-alloy steel, and stainless steel. For corrosive services, cyclones can be lined with soft natural rubber, neoprene, or PVC.

(b) **Construction**

There are several types of cyclones, but a typical single high-efficiency cyclone is shown in Fig. 1.25.

Installation and inspection should be as specified in the manufacturer's manual book.

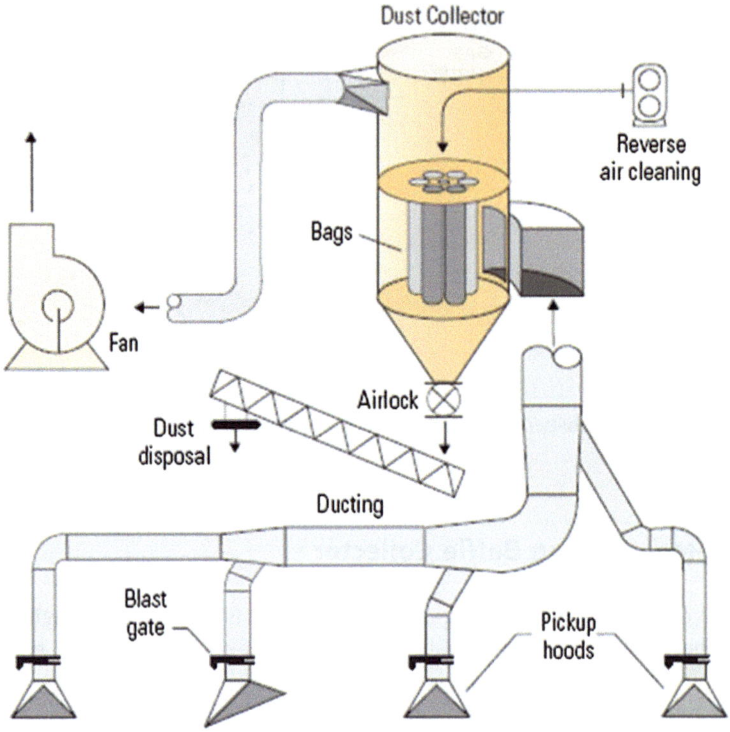

Fig. 1.24 Recirculating baffle collector

1.23 Fabric Filters

Fabric filters are separators in which the gas passes through a porous layer, which retains the particles.

1.23.1 Fibrous Filter Separators

Fibrous filter separators are filtering separators where the particles are separated by means of a medium consisting of natural, mineral, synthetic, or metallic fibers which constitute a woven or an unwoven material. These filtering media are generally in the form of bags or pockets.

A typical diagram of baghouse filter separator is shown in Fig. 1.26.

Installation and inspection should be as specified in the manufacturer's manual book.

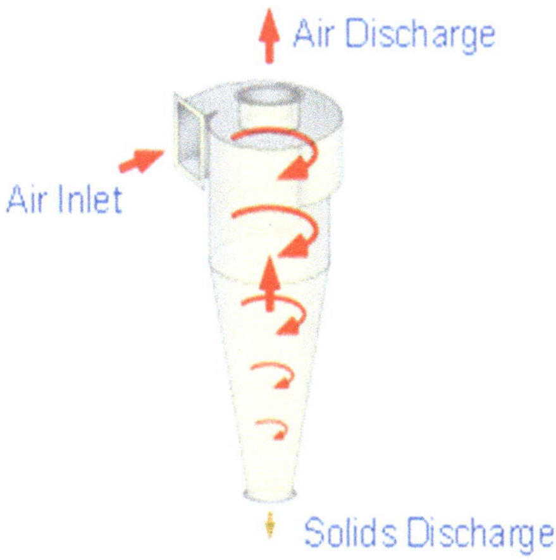

Fig. 1.25 Single involute cyclone

1.24 Wet Scrubbers

Separators in which forces are applied to promote the transfer of particles from a gaseous flow to a liquid phase are later removed from the gaseous flow by other mechanisms.

1.24.1 Bubble Washers or Packed Bed Scrubbers

In this type of scrubber, the dirty gas is passed upward through the tortuous interstices.

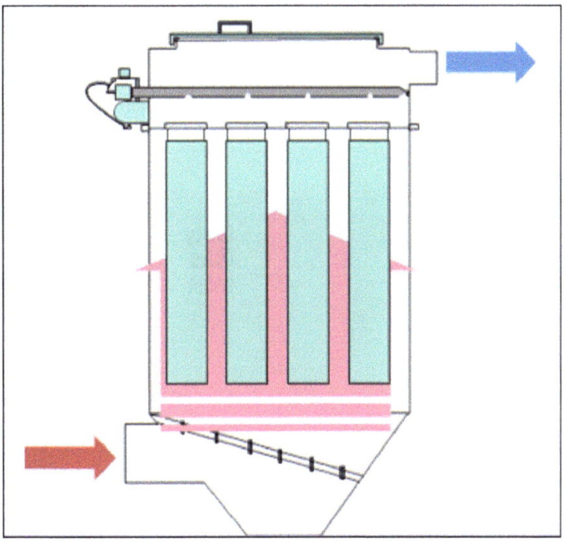

Fig. 1.26 Diagram of baghouse

A great variety of packing could be used including glass, plastic spheres, raschig rings, berl saddles, tellerettes, partitioned rings.

A typical single-bed scrubber is shown in Fig. 1.27.

The packed scrubbing section could be thin, thick, or multiple with redistribution of water between them.

Installation and inspection should be as specified by manufacturers in their manual books.

1.25 Spray Washers

In this type of scrubbers' fine water sprays, wash the gas with water and settle the dust as sludge in the sludge tank (see also ISO 6584).

In general, the body of scrubbers is made from mild steel.

Basic principle of this type scrubber is shown in Fig. 1.28, which includes, dust cycle, demist section, and water cycle.

Captured wetted dust particles settle as droplets under the influence of gravity to the surface of the water pool.

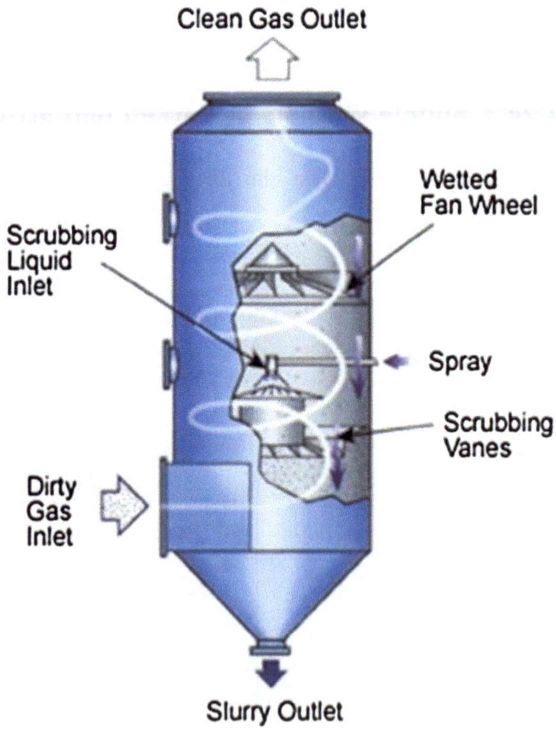

Fig. 1.27 Counter-current flow scrubber

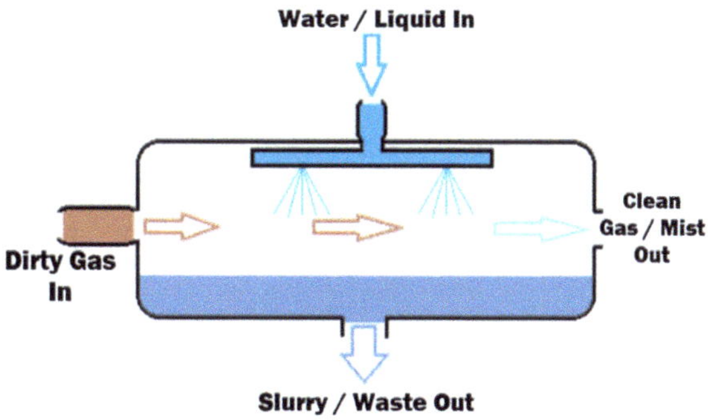

Fig. 1.28 Wetted impingement baffle scrubber

1.25.1 Water Cycle

Water is recycled at a rate of 3.78–37.8 L/min (2–10 gal/min) for every 28.32 m^3 (1,000 ft^3) of gas cleaned.

Installation and inspection should be as specified by manufacturers in their manual books.

1.26 Restricted Flow Scrubbers

Restricted flow scrubbers are wet separators in which the particles are brought into contact with the washing liquid in a restricted zone, which causes a change in pressure or velocity conditions in the flowing gas line Venturi, or orifice scrubbers and induced gas scrubbers (see also ISO 6584).

Construction of vertical downward gas flow Venturi scrubber is shown in Fig. 1.29 in which scrubbing liquid is introduced into the tapered inlet section of the Venturi by overflowing a circular weir. Liquid is fed tangentially into the weir through a pipe, thereby eliminating nozzle plugging. This design could use water containing large quantities of solids.

Installation and inspection should be as specified by manufacturers in their manual books.

1.27 Electric Precipitators

An electric precipitator separates entrained particulate matter from a gas stream by first charging the dust to a negative voltage, precipitating it onto grounded collecting electrodes.

In this type of precipitator, grounded collecting electrodes which electrically insulated should be used. High-voltage wires (50,000 V) could be used for this purpose. The major components consist of

Gas-tight casing including:

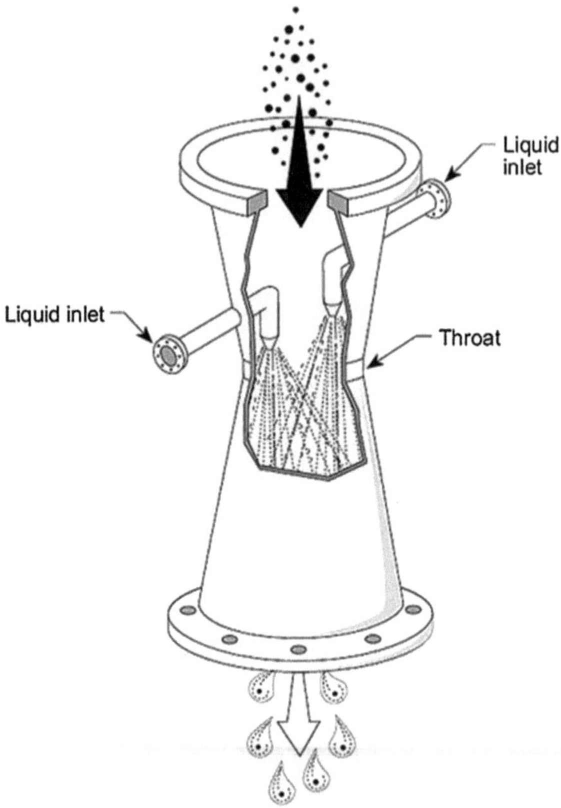

Fig. 1.29 Vertical downward gas flow Venturi scrubber

(a) Hoppers.
(b) High-voltage discharge electrode system.
(c) Grounded collectrode system (cold-rolled steel).
(d) High-voltage supply (silicon diode power packs).

Installation and inspection should be as specified by manufacturers in their manual books.

The basic structure of electric precipitator is gas ionization, dust charging, dust precipitation, dust layer builds, collectrode rapping, and dust fall into hopper.

Installation and inspection should be as specified by manufacturers in their manual books.

1.28 Equipment for Gaseous Waste Disposal

In these sections, equipment suitable for the elimination of gaseous pollutant will be discussed.

1.28.1 Dispersion with Stacks

A simple method for removing pollutants is dispersion. In this method, gas can be dispersed through stack or chimney and is diluted in the air when directly vented into the atmosphere (fog dispersion).

Small-diameter stacks or chimneys are constructed entirely of metal (steel, stainless steel).

1.28.2 Absorption

Absorption can take place either with or without chemical reaction. Gas absorbers for wasting processes consist of five types of apparatus:

(a) Packed columns;
(b) plate columns;
(c) spray towers;
(d) high-energy scrubbers and;
(e) gas–solid absorption.

1.28.3 Packed Columns

Packed columns should be designed of material resistant to the corrosion of the absorbed gas–liquid mixture. Material for packing section could be ceramic metal, or plastic, which provides a large amount of surface area per unit volume.

A distributor should be used to distribute the liquid phase over the packing.

A typical layout of packed column is shown in Fig. 1.30. Recycling in one or more packed section should increase the concentration of the absorbed gas. The gaseous effluent will exit from the top of the tower.

Installation and inspection should be as specified by manufacturers in their manual books.

1.28.4 Plate Columns

Plate columns should be used for the absorption of gases, and vapors when low liquid rate is desirable. Plate columns could be used in a variety type of standards as bubble cap, bubble trays, and sieve trays. Further specifications should be obtained from manufacturers. In the plate column, because of the distribution mechanism, smaller quantities of liquid solvent could be used for the absorption of small quantities of vapor gas solute. A typical layout of plate column is shown in Fig. 1.31.

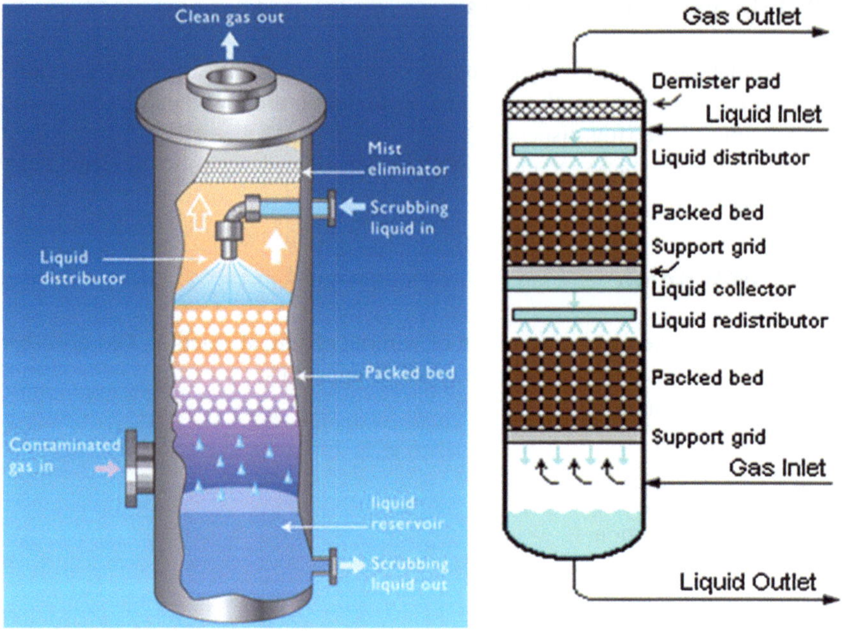

Fig. 1.30 Packed columns

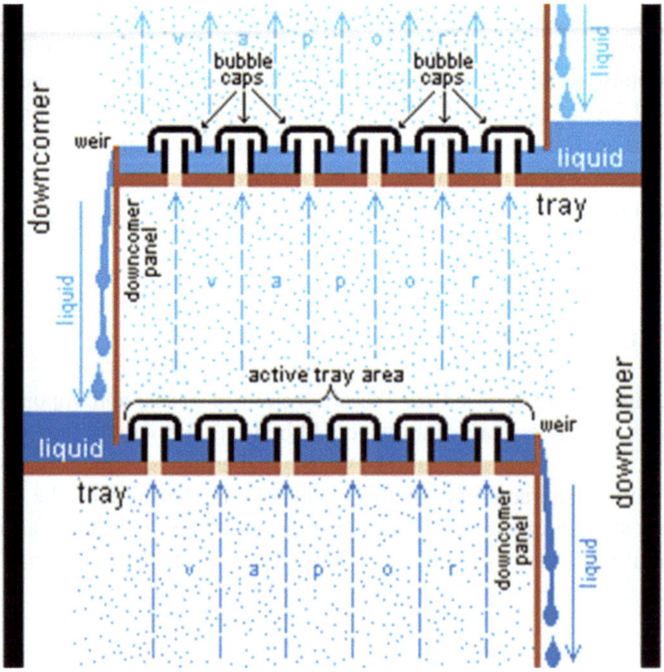

Fig. 1.31 A typical layout of plate column

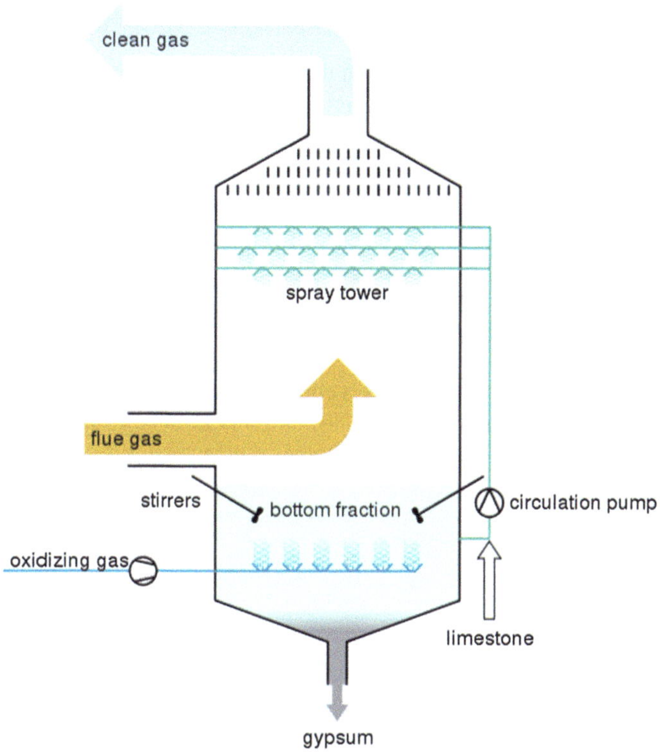

FIG. 1.32 Gas solid absorption

1.29 Gas Solid Absorption

In dealing with high gas flow rates containing substantial quantity of pollutant, gas–solid absorption method should be used.

There are varieties of solid absorbent with different processes for the removal of pollutant gas. Typical layout of processes for the removal of SO_2 is shown in Fig. 1.32 in which SO_2 is removed from the flue gases by a powdery sorbent (lime or limestone) fed into the intake of the desulfurization chamber. The basic reaction is

$$CaO + {}^{1}\!/_{2}O_2 + SO_2 \rightarrow CaSO_4 \tag{1.41}$$

Because of the sorbent's small particle size (1–10 microns) and the optimum temperature 850–1,100 °C, more than 80 % of the sulfur could be captured.

1.30 Condensation

Condensation could be considered when the waste gas has substantial quantities of a condensable material such as air which is saturated with water or other vapor. The equipment, which should be used for this process, are
(a) Tubular surface condenser,
(b) Tubular air-cooled condenser,
(c) Direct-contact condenser.

1.31 Combustion

Destruction of a waste gas or vapor by a combustion process is called incineration. There are three types of incineration:
(a) Direct flame;
(b) Thermal;
(c) Catalytic.

1.31.1 Direct Flame

Direct flame should be used when handling gaseous waste materials that are at or near their lower combustible limit when mixed with air. Also, it could be used when the waste gas itself is a combustible mixture without the addition of air.

The equipment for direct flame incineration should be burner or combustor firing into some enclosure or into the open, which is called flare.

1.31.2 Thermal Incineration

In this method, waste gas should be injected directly through a burner along with auxiliary fuel such as natural gas.

Typical type could be line burner in which a gas pipe with a number of holes injects a fuel such as natural gas into the waste gas stream at the point of ignition. This type could be seen in Fig. 1.33.

Installation and inspection should be as specified by the manufacturers in their manual books.

1.31.3 Catalytic Incineration

Catalytic incineration should be considered for gaseous wastes containing low concentration of combustible materials and air.

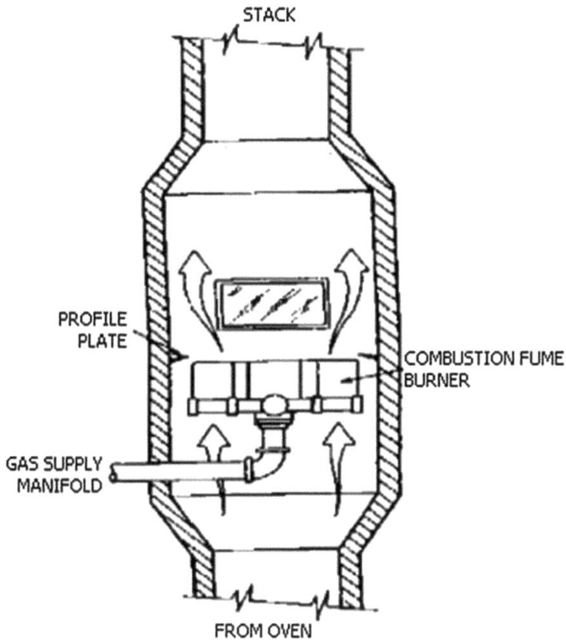

Fig. 1.33 Duct-type fume burner

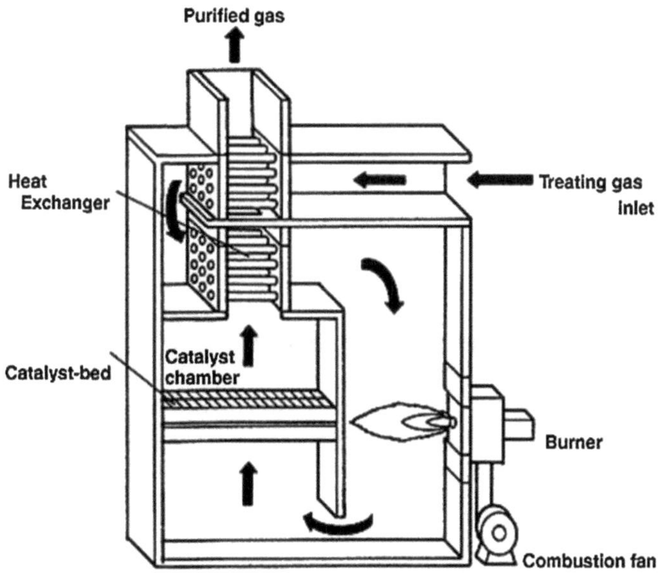

Fig. 1.34 Catalytic incinerator

There are different types of catalysts, but in general, they may be noble metals such as platinum or palladium dispersed on catalyst support which is alumina.

In this method, the gas is preheated to a temperature to cause the reaction to occur on the surface of the catalyst. A typical sketch of equipment is shown in Fig. 1.34.

1.32 Selection of Blowdown Systems

Petroleum industry process units are equipped with a collection unit called the blowdown system. It allows the safe disposal of liquid and vapor hydrocarbons that are vented in pressure relief valves or drawn from the unit. This system can also be used to purge the unit in case of shutdowns. Blowdown materials are partly liquid and partly vapor. The liquid cut is either recycled into the refinery or sent to the wastewater treatment. The vapor cut is either recycled or discharged directly to the atmosphere or flared. When discharged directly to the atmosphere, emissions consist principally in hydrocarbons. When flared, sulfur oxides are emitted. The emission rate of the blowdown system depends on the amount of equipment considered, the frequency of discharges, and the blowdown system controls.

While the various systems for the disposal of voluntary or involuntary vapor or liquid are mentioned below, the actual selection of a disposal system shall be conducted in accordance with the expected frequency, duration of operation, required capacity, and fluid properties.

1.32.1 Blowdown System for Vapor Relief Stream

Systems for the disposal for voluntary and involuntary vapor discharges are

1. To atmosphere.
2. To lower-pressure process vessel or system.
3. To closed pressure relief system and flare.
4. Acid gas flare.

1.32.2 Vapor Discharge to Atmosphere

Vapor relief streams shall be vented directly to atmosphere if all of the following conditions are satisfied (for a complete discussion on the subject, see API RP 521):

1. Such disposal is not in conflict with the present regulations concerning pollution and noise.
2. The vapor is effectively non-toxic and non-corrosive.
3. Vapor which is lighter than air or vapor of any molecular mass that is non-flammable, non-hazardous, and non-condensable.
4. There is no risk of condensation of flammable or corrosive materials.

5. There is no chance of simultaneous release of liquid, apart from water.
6. Relief of flammable hydrocarbons direct to the atmosphere should be restricted to cases where it can be assured that they will be diluted with air to below the lower flammable limit. This should occur well before they can come in contact with any source of ignition.

The above condition can most easily be met if the vapors to be released have a density less than that of air. However, with proper design of the relief vent, adequate dilution with air can be obtained in certain cases with higher-density vapors. Methods of calculation are given in API RP 521 Sect. 4.3

1.32.2.1 Exceptions:

1. Vapor from depressuring valves shall be discharged to a closed pressure relief system.
2. Vapor, which contains 1 % H2S or more by volume, shall be discharged to a closed pressure relief system.

1.32.3 Vapor Discharge to Lower-Pressure Process Vessel or System

Individual safety/relief valves may discharge to a lower-pressure process system or vessel capable of handling the discharge.

Although this type is rarely used, it is effective for discharges that contain materials which must be recovered.

1.32.4 Vapor Discharge to Closed Pressure Relief System and Flare

Flaring is a safety measure used in petroleum industries to ensure that gases are safely disposed of. A flare is a device that burns hydrocarbons emitted from emergency process vents or pressure relief valves. It is usually assumed that flares have a combustion efficiency of at least 98 %.

The combustion reaction is

$$C_xH_y + (x + y/4)O_2 \rightarrow x\, CO_2 + y/2\, H_2O \tag{1.42}$$

Principally, carbon dioxide is emitted from flares, but also organic compounds and carbon monoxide, NOx, SOx, and soot. It is actually impossible to estimate flare emissions; however, several measures can be adopted to minimize these emissions:

- Use of efficient flare tips and optimization of the size and number of burning nozzles.

- Maximization of flare combustion efficiency by controlling and optimizing flare fuel/air/steam ratio.
- Minimization of flaring from purge without compromising safety, through measures such as purge gas reduction devices, flare gas recovery units, inert purge gas.
- Installation of high integrity instrument pressure protection systems, where appropriate, to reduce over pressure events and avoid or reduce flaring situations.
- Minimization of liquid entrainment in the gas flare stream with a suitable liquid separation system.
- Implementation of burner maintenance and replacement programs to ensure continuous maximum flare efficiency.

In all cases where the atmospheric discharge or release of vapor to a lower-pressure system is not permissible or practicable, vapor shall be collected in a closed pressure relief system which terminates in a flare, namely flare system. Where the concentration of H_2S is such that condensation of acid gas is probable, provision for a separate line, heat-traced, shall be considered. In all cases, the installation of a closed pressure relief system shall result in a minimum of air pollution and the release of combustion products.

1.33 Acid Gas Flare

In process plants where H_2S free and H_2S containing stream are to be flared, consideration should be given to the installation of a separate header and flare stack assembly for the H_2S containing streams. The following provisions should be studied for the acid gas flare assembly:

1. Automatic injection of fuel gas downstream of H_2S pot in order to make the combustion stable.
2. Steam injection for smokeless operation shall not be considered for H_2S flare tip.
3. A common pilot igniter shall be used to ignite all flare stacks including the acid flare.
4. The H_2S flare header and subheaders may be heat-traced in order to prevent the condensation acid gas.

1.34 Design of Disposal System Components

Depending on the process plant under consideration, a disposal system could consist of a combination of the following items: piping, knockout drum, quench drum, seal drum, flare stack, ignition system, flare tip, and burning pit.

1.34.1 Piping

In general, the design of disposal piping should conform to the requirements of ANSI/ASME B31.3. Installation details should conform to those specified in API Recommended Practice 520, Part II.

The design of inlet piping should be in accordance with API RP 521, Sect. 5.4.1.

The sizing should be in accordance with API RP 521, Sect. 5.4.1.2, and API RP 520, Part I, Section 7.

The design should be in accordance with API RP 521. The design should be in accordance with API RP 521 and ANSI/ASME B31.3.

1.34.2 Drainage

Disposal system piping should be self-draining toward the discharge end. Pocketing of discharge lines should be avoided. Where pressure relief valves handle viscous materials or materials that can solidify as heat is cooled to ambient temperature, the discharge line should be heat-traced. A small drain pot or drip leg may be necessary at low points in lines that cannot be sloped continuously to the knockout or blowdown drum. The use of traps or other devices with operating mechanisms should be avoided.

1.34.3 Details

1. **Safety/relief valve connection to the header**

   Normally, the laterals from individual relieving devices should enter a header from above.

2. **Safety/relief valves connection when installed below the relief header**

   Laterals leading from individual valves located at an elevation above the header should drain to the header. Locating a safety valve below the header elevation in closed systems should be avoided. Laterals from individual valves that must be located below the header should be arranged to rise continuously to the top of the header entry point. However, means should be provided to prevent liquid accumulation on the discharge side of these valves. In this regard, the following should be taken into consideration:

   For the branch header which must be connected to the main header from a lower level than the main header, e.g., sleeper flare piping, a drain pot must be installed.

   If a safety/relief valve must be installed below the flare header, the outlet line leading to the flare header shall be heat-traced from the safety/relief valve to their highest point. But the arrangement of safety/relief valve must be reviewed, as such; an arrangement is not permitted for safety/relief valves which discharge a medium which can leave a residue. The heat-tracing can be omitted if the safety/relief valve in question handles only products which vaporize completely, or do not condense at all, at the lowest ambient temperature.

3. **Purge Point of Gas for Dry Seal**
   (a) A continuous fuel gas purge shall be installed at the end of the main header and the end of any major subheader. The fuel gas purge shall be controlled by means of a restriction orifice.
   (b) Purge gas volume shall be determined such that a positive pressure is maintained and air ingress is prevented.
4. **Insulation of Flare Line**
   Normally, insulation of flare line (including outlet line of safety/relief valve) is not required except for personnel protection.
   But to avoid hydrate formation or ice accumulation, etc., within the flare line, the use of insulation or heat-tracing shall be considered.
5. **Location of Safety/Relief Valve**
   More than one piece of equipment may be protected by a common safety/relief valve, provided they are connected by a line of sufficient size and that no block valve exists on the connecting lines.
6. **Valves on Inlet/Outlet Line of Safety/Relief Valve**
   Unless otherwise specified by the company, all safety relief valves must have block valves on the inlet and outlet to facilitate maintenance. The block valves must be full bore and locked open. Safety valves discharging to the atmosphere shall not have block valves on the outlet. A bypass line with a valve shall be provided for each safety valve.
7. **Provision for Installation of Drain Holes**
   Where individual valves are vented to the atmosphere, an adequate drain hole (a nominal pipe size of DN 15 is usually considered suitable) should be provided at the low point to ensure that no liquid collects downstream of the valve. The vapor flow that occurs through this hole during venting is not generally considered significant, but each case should be checked to see whether the drain connection should be piped to a safe location. Vapors escaping from the drain hole must not be allowed to impinge against the vessel shell, since accidental ignition of such vent streams can seriously weaken the shell.
8. **Angle Entry Into the Relief Header**
   The use of angle entry, an entry at 45° (0.79 rad) or even 30° (0.52 rad) to the header axis for laterals, is much more common in relieving systems than in most process piping systems.
9. **Installation of Valves and Blinds in Relief Headers**
   Means (valve and blind) must be provided to isolate each unit from the flare system for safety and maintenance.
   Extreme caution must be exercised in their use to ensure that equipment which is operating is not isolated from its relieving system. Valves in the header system, if used, should be mounted so that they cannot fail in the closed position (for example, a gate falling into its closed position).

10. **Slope of Flare Header**
    A slope of 1 m in 500 m is suggested for the flare header.
11. **Absorption of Thermal Expansion in Headers by Looped Pipes**
    (a) As a rule, headers shall be designed so that thermal expansion generated in headers can be absorbed by the bent parts of the headers. In other words, the piping route of headers shall incorporate several bends.
    (b) If thermal expansion cannot be absorbed by the above method, absorption by looped pipes shall be considered. Looped parts shall have no drain pocket.
12. **Absorption of Thermal Expansion by Expansion Joints**
    (a) As a rule, no expansion joints shall be used. The use of expansion joints is limited to the case in which thermal expansion cannot be absorbed by pipes alone because of a short route, e.g., the route between the seal drum (or knockout drum) and the flare stack.
    (b) Drain pipes shall be installed at bellows or other concave parts where drain is likely to remain.
    (c) The conditions for selecting bellows (design condition, materials) shall be specified clearly.
13. **Solids Formation**

The possibility of solids forming within the disposal system must be studied considering all related aspects, such as hydrate formation, water or heavy hydrocarbon presence, autorefrigeration. Consideration should be given to separate disposal system so that the possibility of solid formation is eliminated.

1.35 Quench Drum

A quench drum is provided as a means of preventing liquid hydrocarbon condensation in the flare system, to reduce flare capacity requirements or to prevent discharge of condensable hydrocarbons to the atmosphere. In some cases, it serves the additional purpose of reducing the maximum temperature of flare gases and hence minimizing thermal expansion problems in the mechanical design of flare headers.

The quench drum functions by means of direct-contact water spray arrangement which condenses entering heavy hydrocarbon vapors. Condensed hydrocarbons and effluent water are discharged through a seal to the sewer or pompous to slop tankage. On the other hand, uncondensed hydrocarbon vapors are vented to the flare or to the atmosphere. Figure 1.35 presents a typical quench drum.

1.35.1 Details

(a) The quench drum shall have a design pressure capable of withstanding the maximum back pressure. Minimum design pressure is 350 kPa gage.

Fig. 1.35 Typical quench drum

(b) Water requirements are normally based on reducing gas and liquid outlet temperatures to about 50 °C. Selection of the optimum temperature is based on considerations of temperature and composition of entering streams, and the extents to which subsequent condensation of effluent vapors downstream of the drum can be tolerated.
It is generally assumed that no more than 40–50 % of the liquid fed will be vaporized. The water supply should be taken from a reliable water system. If a recirculating cooling water system is used, then the circulating pumps and cooling water basin must have adequate capacity to supply the maximum quench drum requirements for 20 min.
The seal height in the liquid effluent line (assuming 100 % water) is sized for 175 % of the maximum operating pressure, or 3 m, whichever is greater.

(c) Should the quenched hydrocarbons be of a sour nature; Provisions shall be made for proper disposal system and due consideration be given to material specification.

1.36 Sizing Knockout Drum

Sizing a knockout drum is generally a trial-and-error process. First, the drum size required for liquid entrainment separation is determined. Liquid particles will separate when the residence time of the vapor or gas is equal to or greater than the time required to travel the available vertical height at the dropout velocity of the liquid particles, and the vertical gas velocity is sufficiently low to permit the liquid droplet to fall. This vertical height is usually taken as the distance from the liquid surface.

The vertical velocity of the vapor and gas must be low enough to prevent large slugs of liquid from entering the flare. Since the flare can handle small-sized liquid droplets, the allowable vertical velocity in the drum may be based on that necessary to separate droplets from 300 to 600 μm in diameter. The dropout velocity of a particle in a stream is calculated as follows:

$$U_c = 1.15\sqrt{\frac{gD(\rho_L - \rho_V)}{\rho_V(C)}} \tag{1.43}$$

C Drag coefficient.
D Droplet diameter, m.
g Acceleration due to gravity, in (9.8 m/s^2).
U_c Particle dropout velocity, in (m/s).
ρ_L Liquid density, kg/m^3.
ρ_V Vapor density, kg/m^3.

This basic equation is widely accepted for all forms of entrainment separation.

The second step in sizing a knockout drum is to consider the effect any liquid contained in the drum may have on reducing the volume available for vapor/liquid disengagement. This liquid may result from (1) condensate that separates during a vapor release or (2) liquid streams that accompany a vapor release. It is suggested that the volume occupied by the liquid be based on a release lasting 20–30 min.

Any accumulation of liquid retained from a prior release (pressure relief valves or other sources) must be added to the liquid indicated in Items 1 and 2 above to determine the available vapor disengaging space. However, for situations where the knockout drum is used to contain large liquid dumps from pressure relief valves on other sources where there is not significant flashing and the liquid can be removed promptly, it would not usually be necessary to consider these volumes relative to vapor disengaging.

The economics of vessel design should be considered when selecting a drum size and may influence the choice between a horizontal drum and a vertical drum. When large liquid storage is desired and the vapor flow is high, a horizontal drum is often more economical. Split entry of exit decreases the size of the drum for large flows.

As a rule, drum diameters over 3.3 m should apply split-flow arrangements for best economics. Horizontal and vertical knockout drums are available in many designs, the main differences consisting in how the path of the vapor is directed. The various designs include the following:

1. A horizontal drum with the vapor entering one end of the vessel and exiting at the top of the opposite end (no internal baffling).
2. A vertical drum with the vapor inlet nozzle on a diameter of the vessel and the outlet nozzle at the top of the vessel's vertical axis. The inlet stream should be baffled to direct the flow downward.
3. A vertical vessel with a tangential nozzle.
4. A horizontal drum with the vapor entering at each end on the horizontal axis and a center outlet.
5. A horizontal drum with the vapor entering in the center and exiting at each end on the horizontal axis.
6. A combination of a vertical drum in the base of the flare stack and a horizontal drum upstream to remove the bulk of the liquid entrained in the vapor. This combination permits the use of larger values for the numerical constant in the velocity equation.

The following sample calculations have been limited to the simplest of the designs, Items 1 and 2. The calculations for Items 4 and 5 would be similar, with one-half of the flow rate determining one-half of the vessel length.

The normal calculations would be used for Item 3 and will not be duplicated here.

Assume the following conditions: A single contingency results in the flow of 25.2 kilograms per second of a fluid with a liquid density of 496.6 kilograms per cubic meter and a vapor density of 2.9 kilograms per cubic meter, both at flowing conditions. The pressure is 13.8 kPa gage, and the temperature is 149 °C. The viscosity of the vapor is 0.01 centipoise.

Also, the fluid equilibrium results in 3.9 kilograms per second of liquid and 21.3 kilograms per second of vapor.

In addition, 1.89 cubic meters of storage for miscellaneous drainings from the units is desired. The droplet size selected as allowable is 300 μm in diameter. The vapor rate is determined as follows:

$$vapor\ rate = \frac{21.3\,kg/S}{2.9\,kg/m^3} = 7.34\,m^3/S \tag{1.44}$$

The drag coefficient, C, is determined from Fig. 1.36 as follows:

$$C(\mathrm{Re})^2 = \frac{0.13 \times 10^8 (2.9)(300 \times 10^{-6})^3 (496.6 - 2.9)}{(0.01)^2} = 5025 \tag{1.45}$$

From Fig. 1.36, $C = 1.3$. The dropout velocity, U_c, is calculated as follows:

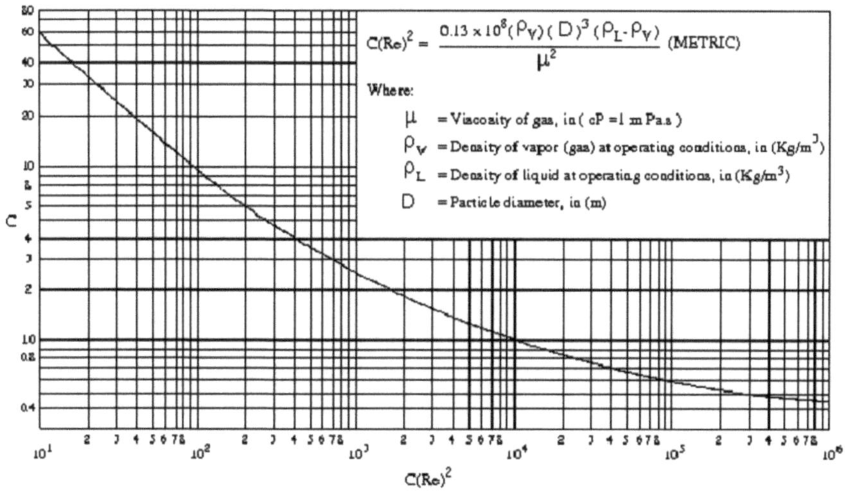

Fig. 1.36 Determination of drag coefficient

$$U_c = 1.15\left[\frac{(9.8)(300\times10^6)(496.6-2.9)}{(2.9)(1.3)}\right]^{0.5} = 0.71 \tag{1.46}$$

Assume a horizontal vessel with an inside diameter, D_i, and a cylindrical length, L. This gives the following total cross-sectional area:

$$\mathrm{A_t} = \frac{\pi}{4}(Di)^2 \tag{1.47}$$

where

A_t Total vessel cross-sectional area, in (m^2).
D_i Inside diameter of vessel, in (m).

Liquid holdup for 30 min release from the single contingency, in addition to the slop and drain volume, is desired. The volume in the heads is neglected for simplicity. The liquid holdup required is therefore calculated as follows:

1. The slop and drain volume of 1.89 cubic meters will occupy a bottom segment as follows:

$$A_{L1}(Square\ meters) = \left(\frac{1.89\,m^3}{L}\right) \tag{1.48}$$

2. A total of 3.9 kilograms per second of condensed liquids with a density of 496.6 kilograms per cubic meter accumulated for 30 min will occupy a cross-sectional segment (see above) as follows:

$$A_{L2} = (\frac{3.9\,kg/S}{496.6\,kg/m^3})(\frac{60\,S}{1\,\text{min.}})(30\,\text{min})(\frac{1}{L, meter}) \tag{1.49}$$

The cross-sectional area remaining for the vapor flow is as follows:

$$A_V = A_t - (A_{l1} + A_{l2}) \tag{1.50}$$

where

A_{L1} Vessel segment area occupied by slops and drains, in (m^2).
A_{L2} Vessel segment area occupied by condensed liquid, in (m^2).
A_V Vessel cross-section area available for vapor flow, in (m^2).

The vertical depths of the liquid and vapor spaces are determined using standard geometry, where h_{L1} = depth of slops and drains, $h_{L1} + h_{L2}$ = depth of all liquid accumulation, and h_v = remaining vertical space for the vapor flow. The total drum diameter, h_t, is calculated as follows:

$$h_t = h_{L1} + h_{L2} + h_v \tag{1.51}$$

The adequacy of the vapor space is verified as follows. The vertical drop available for liquid dropout is equal to h_v. The liquid dropout time is determined as follows:

$$\theta = \left(\frac{1}{U_c(meter\, per\, second)}\right)\left(\frac{h_V}{100\ centimeter\, per\, meter}\right) \tag{1.52}$$

The drum length required is determined as follows:

$$\boldsymbol{L_{min}} = (\boldsymbol{U_v\ meters\ per\ second})(\theta\ \boldsymbol{seconds}) \tag{1.53}$$

L_{min} must be less than or equal to the above assumed cylindrical drum length, L; otherwise, the calculation must be repeated with a newly assumed cylindrical drum length.

Table 1.30 summarizes the calculations above for horizontal drums with various inside diameters to determine the most economical drum size. Drum diameters in 15-cm increments are assumed, in accordance with standard head sizes.

1.37 Flares

Flare systems provide for the safe disposal of gaseous refinery wastes. Depending on local environmental constraints, these systems can be used for

1. Extensive venting during start-up or shutdown.

Table 1.30 Optimizing the size of horizontal knockout drum

Trial no.	Assumed drum inside diameter Di (m)	Assumed drum cylindrical length, L (m)	Cross-sectional area (m^2)				Vertical depth of liquid and vapor spaces (cm)				Liquid dropout time, q (s)	Vapor velocity, U v (m/s)	Required drum length, L min (M)
			A1	AL1	AL2	Av	hl1	hl1 +hl2	h v	d i			
1	22.4	5.79	4.67	0.33	2.45	1.9	30	140	104	224	1.45	3.9	5.6
2	2.29	6.25	4.1	0.3	2.27	1.53	29	137	91	229	1.28	4.8	6.2
3	2.13	6.86	3.57	0.28	2.07	1.23	28	133	81	213	1.13	6	6.7
4	1.98	7.62	3.08	0.25	1.86	0.98	27	128	70	198	0.98	7	7.4

2. Venting of excess process plant gas.
3. Handling emergency releases from safety valves, blowdown, and depressuring systems.

Designs will vary considerably, depending upon the type of connected equipment and the complexity of the overall system. A flare system generally consists of an elevated stack, means to maintain burning conditions at the top of stack, and means to prevent flashback within the system.

1.37.1 Sizing

The sizing of flares requires determination of the required stack diameter and the required stack height.

Since the flare tip is open to the atmosphere, high gas velocities are expected at this point. Very high tip velocities cause a phenomenon known as blow-off where the flame front is lifted and could eventually turn into a blowout. Very low velocities could damage the flare tip due to high heat intensities and smoking. In this case, ingress of air in the system and creation of a flammable mixture are possible. Therefore, determination of the right flare diameter is important as far as operation of the system is concerned.

The location and height of flare stacks should be based on the heat release potential of a flare, the possibility of personnel exposure during flaring, and the exposure of surrounding plant equipment. There are exposure limitations set forth which must be taken into consideration. This in effect fixes the distance between the flame and the object. Now, if there are limitations on the location (distance), then the stack height can be calculated; otherwise, an optimum trade-off between height and distance should be applied.

Wind velocity, by tilting the flame in effect, changes the flame distance and heat intensity. Therefore, its effect should be considered in determining the stack height.

If the flare is blown out (extinguished), or if there are environmental hazards associated with the flare output, the possibility of a hazardous situation downwind should be analyzed.

1.37.1.1 (a) Diameter

Flare stack diameter is generally sized on a velocity basis, although pressure drop should be checked. Depending on the volume ratio of maximum conceivable flare flow to anticipated average flare flow, the probable timing, frequency, and duration of those flows, and the design criteria established for the project to stabilize flare burning, it may be desirable to permit a velocity of up to 0.5 Mach for a peak, short-term, infrequent flow, with 0.2 Mach maintained for the more normal and possibly more frequent conditions. Smokeless flares should be sized for the conditions under which they are to operate smokelessly.

Pressure drops as large as 14 kPa have been satisfactorily used at the flare tip. Too low a tip velocity can cause heat and corrosion damage. The burning of the gases becomes quite slow, and the flame is greatly influenced by the wind. The low-pressure area on the downwind side of the stack may cause the burning gases to be drawn down along the stack for 3 m or more. Under these conditions, corrosive materials in the stack gases may attack the stack metal at an accelerated rate, even though the top 2.4–3 m of the flare is usually made of corrosion-resistant material.

1.37.2 Design Details

1. **Smokeless Flares**
   Smoke-free operation of flares can be achieved by various methods, including steam injection, injection of high-pressure waste gas, forced draft air, operation of flares as a premixed burner, or distribution of the flow through many small burners. The most common type of smokeless flare involves steam injection.
   The assist medium mass requirements are low for steam and fuel gas because of their high velocity relative to the flare gas. Typical values for steam or fuel gas are from 0.20 to 0.50 kg of assist gas per kilograms of hydrocarbon flow.
2. **Flashback Protection**
   The most common method of preventing propagation of flame into the flare system due to entry of air is the installation of a seal drum. Flame arresters are occasionally used for flashback protection; however, they are subject to plugging, and their application is limited.
   Alternatively, continuous introduction of purge gas can be used to prevent flash back. A safe condition exists in situations involving hydrocarbon air mixtures if a positive flow of oxygen-free gas is maintained, allowing the oxygen concentration to be no greater than 6 % at a point 7.6 m from the flare tip. The injection rate should be controlled by a fixed orifice to ensure that supply remains constant and is not subject to instrument malfunction or maladjustment. Molecular seals can be used to minimize purge gas rates.
3. **Ignition**
   To ensure ignition of flare gases, continuous pilot with a means of remote ignition are recommended for all flares. The most commonly used type of igniter is the flame-front propagation type, which uses a spark from a remote location to ignite a flammable mixture.
   Pilot igniter controls are located at a safe distance from the base of elevated flares and at least 30 m from ground flares.
   It is recommended that a low-pressure alarm for the pilot gas be provided so that the operator in the control room becomes aware of pilot blowout.
   Reliable pilot operation under all wind and weather conditions is essential. Flaring operations are for the most part intermittent and non-scheduled. The

flare must be instantly available for full emergency duty to prevent any possibility of a hazardous or environmentally offensive discharge to the atmosphere. Wind shields and flame retention devices may be used to ensure continuous piloting under the most adverse conditions.

4. **Fuel System**
   Fuel gas supply to the pilots and igniters must be highly reliable. Since normal plant fuel sources may be upset or lost, it is desirable to provide a backup system connected to the most reliable alternative fuel source, with provision for automatic cut-in on low pressure. Use of waste gas with low energy content or with unusual burning characteristic should be avoided. Parallel instrumentation for pressure reduction is frequently justifiable. The flare fuel system should be carefully checked to ensure that hydrates cannot present a problem. Because of small lines and long-exposed runs, large vertical rises up the stack, and pressure reductions, the use of a liquid knockout pot or scrubber after the last pressure reduction is frequently warranted. If at all feasible in terms of distance, relative location, and cost, it is considered good practice to install a low-pressure alarm on the fuel supply after the last regulator or control valve so that operators will be warned of any loss of fuel to the pilots.

5. **Fired or Endothermic Flares**
   When low-heating-value gases are to be sent to a flare stack, fired or endo thermic flare is used (sulfur plant tail gas presents an example).
   Generally, if the heating value of the gas to be flared is less than 4,280 kJ/m^3, a fired flare with a high-heating-value assist gas may be required for complete combustion.

6. **Location**
   The location of flares in the vicinity of tall refinery equipment should be examined. Flames or hot combustion products can be carried by the wind, which could cause problems and create hazards to personnel working on these elevated structures at the time of a flare release. As discussed in the section on sizing, flare height and distance are dependent on radiation intensity. When either the height or the distance from the plant of a flare is fixed, the other can be determined. Usually, there are constraints on the distance; therefore, stack height is calculated.
   If there are no constraints on the distance and flare height to be determined, the following guideline is recommended.
   For stack heights less than 23 m, a distance of 91 m is considered, and for stack heights greater than 23 m, a distance of 61 m from the plant is considered.

7. Due consideration should be given to installation of flow measuring equipment on the flare system. Specifically, subheaders handling continuous relief loads from individual units shall be provided with proper flow elements.

1.38 Sample Flare Design Calculations

Given below are the sample data from the documents which are used for calculation purposes.

- Hydrocarbon smokeless vapor flow rate: 7.23 kg/s
- Hydrocarbon peak flow rate: 35.23 kg/s
- Average molecular mass of vapor: 16.23
- Flowing temperature: 312.13 K
- Ratio of specific heats, k = cp/cv: 1.3
- Flowing pressure at flare tip: 103 kPa (absolute)
- Design wind velocity: 85 m/s
- Relative humidity: 100 %.

The data are used for assessing the viability of a flare tip with the air assist system. Initially, the study will be directed to check the flare tip size based on the available standards, followed by the estimation of excess air requirements for ensuring complete combustion in flare tip.

1.38.1 Flare Tip Diameter

Flare tip diameter is generally sized on a velocity basis, although pressure drop should be checked. Depending on the volume ratio of maximum conceivable flare flow to anticipated average flare flow, the probable timing, frequency, duration of those flows, and the design criteria established for the project to stabilize flare burning, it may be desirable to permit a velocity of up to 0.5 Mach for a peak, short-term, infrequent flow, with 0.2 Mach maintained for the more normal and possibly more frequent conditions.

1.38.1.1 Equations Used for Sizing

Diameter based on flare tip velocity:

$$Mach = 1.702 \times 10^{-5}\left(\frac{W}{P_t d^2}\right)\sqrt{\frac{T}{kM}} \tag{1.54}$$

In metric units,

$$Mach = 11.61 \times 10^{-2}\left(\frac{W}{P_t d^2}\right)\sqrt{\frac{T}{kM}} \tag{1.55}$$

Solve for tip diameter:

$$d = \sqrt{11.61 \times 10^{-2} \times W \times \frac{\sqrt{\frac{T}{kM}}}{Mach \times P_t}} \tag{1.56}$$

where

d	flare tip diameter, ft (m) (end or smallest diameter)
k	Ratio of specific heats, Cp/Cv for the vapor being relieved
M	Molecular mass (weight) of the vapor or gas
Mach	ratio of vapor velocity to sonic velocity in vapor, dimensionless. Mach = 0.5 for peak for short-term flow and 0.2 for more normal and frequent conditions
P_t	pressure of the vapor just inside flare tips (at top), psia. For atmospheric release, $P_t = 14.7$ psia [101.3 kPa(abs)]
Pf	Pressure of the vapor just inside the flare tip, in Psi(abs) or (kPa absolute)
T	Temperature of the vapor just inside the flare tip in °R or (K)
W	Vapor relief rate, in lb/hr or (kg/s)

A peak velocity through the flare end (tip) of as much as 0.5 Mach is generally considered a peak, short term. A more normal steady-state velocity of 0.2 Mach is for normal conditions and prevents flare/lift off.

Another similar equation available in the literature yields close results which are generally used for smokeless flares:

$$d_t^2 = (W/1370)\sqrt{\frac{T}{M}} \tag{1.57}$$

where

d_t	flare tip diameter, in.
W	gas vent rate, lb/h
T	gas temperature in stack, ° R
M	molecular weight of gas/vapor.

Based on Mach 0.2 limitation velocity, with ***k*** = *Cp/Cv*, = 0.2, close results can be obtained.

1.38.1.2 Calculation Results for Flare Tip

Table 1.31 presents the summary of results obtained based on the correlations reported in the previous section. Flare tip diameter can also be readily estimated as given below.

With 0.2 Mach number, limitation velocity and following data:

Table 1.31 Calculated flare tip diameter

Mach number	Mass flow rate (Kg/s)	Calculated flare tip diameter (inches)
0.2	7.23 (normal, smokeless)	15.58
0.2	35.23 (peak flow)	34.405
0.5	35.23 (peak flow)	21.76

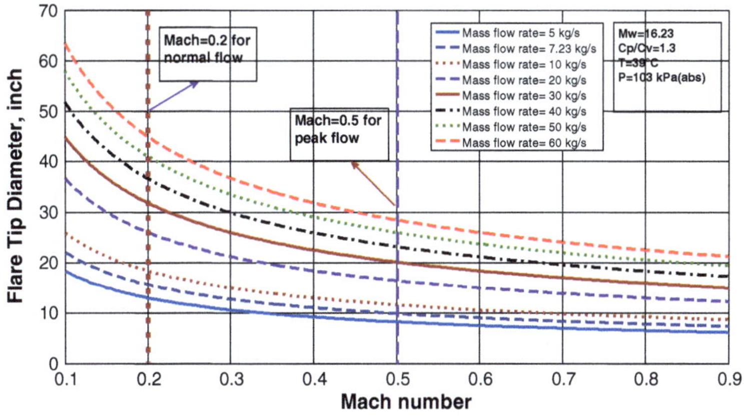

Fig. 1.37 Calculated tip diameter as a function of Mach number

$$k\,\text{ratioof Cp/Cv}) = 0.2$$
$$\text{W} = 35.23\ \text{kg/S} = 279350\ \text{lb/h} \quad \text{(Emergency flow rate)}$$
$$\text{W} = 7.23\ \text{kg/S} = 57330\ \text{lb/h} \quad \text{(Smokeless flow rate)}$$
$$\text{T} = 312.15\ \text{K} = 562.2^\circ\text{R}$$

Estimated diameters are

$$d = 34.14\ \text{in (Emergency flow rate)}$$
$$d = 15.7\ \text{in (for Smokeless flow rate)}$$

Figures 1.37 and 1.38 show the calculated flare tip diameter at various Mach numbers and gas flow rates according to Eqs. 2 and 3. According to the

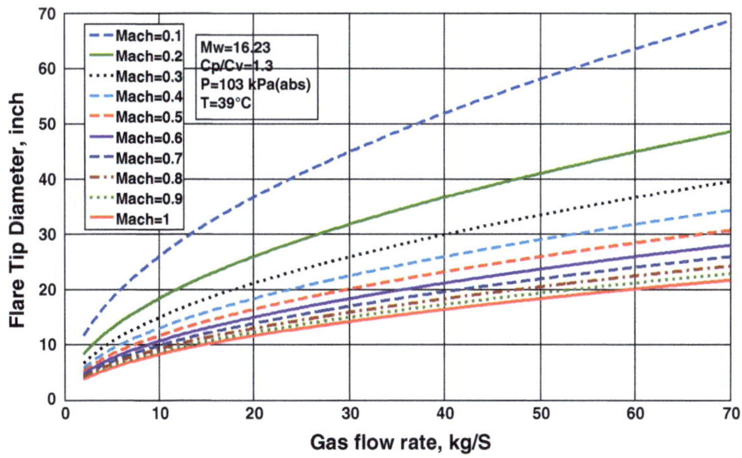

Fig. 1.38 Calculated tip diameter as a function of gas flow rate

information available in the literature, a peak velocity through the flare end (tip) of as much as 0.5 Mach is generally considered for a peak and short-term flow. A more normal steady-state velocity of 0.2 Mach is for normal conditions and prevents flare/lift off. According to API RP 521 standard and reference 2 in this report, Mach = 0.5 to be used for peak and short-term flow and 0.2 for more normal and frequent conditions. On the basis of given data, Table 1 shows that the *maximum* flare tip diameter (peak flow and Mach = 0.2) will be 35 in.

1.39 Organic Emissions

There is a limited amount of information on VOC speciation from refinery process sources. Data that have been found in the literature are presented in Table 1.32. This information is combined with total VOC estimates (i.e., as derived using Table 8) using the following equation:

$$\text{Organic Compound (kg/h)} = [\text{VOC Emission (kg/h)}] \times [(\text{ Speciation Factor, wt\%})/100]$$

Trace Elements and Inorganics

The information on emissions of trace elements and inorganics from refinery process sources is limited, with the only published data relating to emissions from the following sources:

- catalytic cracking units (i.e., both FCCUs and MBCCUs);
- fluid coking operations; and
- sulfur recovery units (SRUs).

Table 1.32 VOC speciation data for refinery process sources

Compound	Weight % of VOCs released from each refinery process operation					
	Catalytic cracking units and includes both FCCUs and moving bed CCUs	Fluid coking	Blowdown systems	Vacuum distillation unit (VDU) condensers	Sulfur recovery units (SRU).	Flare
n-Hexane	NA	3.86	3.86	3.86	NA	NA
Cyclohexane	NA	0.08	0.08	0.08	NA	NA
Formaldehyde	51	8.88	8.88	8.88	4.12	0.817
Acetaldehyde	NA	NA	NA	NA	0.67	0.082
Xylenes	NA	0.19	0.19	0.19	NA	0.041
Benzene	NA	0.38	0.38	0.38	NA	0.083
Toluene	NA	0.44	0.44	0.44	NA	0.041
PAHs	0.15	NA	NA	NA	NA	0.02

Table 1.33 Total particulate emission factors for refinery process sources

Refinery process operation	Total particulate emission factor (kg/m^3 feed to the unit)
Fluidized catalytic cracking units—uncontrolled	0.695
Fluidized catalytic cracking units (ESP & CO boiler)	0.128
Moving bed catalytic cracking units (MBCCU)	0.049
Fluid coking unit—uncontrolled	1.5
Fluid coking unit—ESP & CO boiler	0.0196

However, it is important to note that other sources may release metals and other compounds, although no data are currently available to quantify these. In the absence of actual stack test data, the emission factors presented in Table 1.33 are available to estimate TSP emissions from the refinery sources mentioned above.

When total particulate emissions have been estimated, speciation is performed using the weight fractions presented in Table 1.34, combined with the following equation:

$$ER_i = TSP(WP_i/100) \tag{1.58}$$

where

ER_i The emission rate of the speciated compound "i" from the relevant source (kg/h);

TSP The *total suspended particulates* estimate derived using the emission factors in Table 10 (kg/h). The exception is the carbon disulfide speciation data, as this must be applied to the total VOC estimate from SRUs;

WP_i The appropriate weight percent of species "i" provided in Table 1.34.

1.40 Process Fugitives

Process fugitives at a petroleum refinery typically include the following sources:

- valves;
- flanges;
- pumps;
- connectors;
- compressors; and
- drains.

Although the release from each individual source may be small because of the large number of such sources in a refinery, the total emissions from these sources can be significant.

Emissions from process fugitives are estimated using the following two steps:

1. Total VOC emissions are calculated and

Table 1.34 Trace element and inorganics speciation data for refinery process sources

Metal (& compounds)	CCU (weight % of total particulate matter)		Fluid coking (weight % of total particulate matter)	SRU (weight % of VOC)
	Uncontrolled	Controlled	Uncontrolled	No data available
Manganese	0.022	No data available	0.004	No data available
Nickel	0.088	0.031	0.038	No data available
Copper	0.02	0.003	0.001	No data available
Zinc	0.017	0.006	0.003	No data available
Arsenic	0.002	No data available	0.144	No data available
Selenium	0.002	0.003	0.002	No data available
Antimony	0.035	0.002	0.005	No data available
Lead	0.046	0.01	0.003	No data available
Cobalt	0.002	No data available	No data available	No data available
Cadmium	0.009	0.002	No data available	No data available
Mercury	0.01	0.001	0.002	No data available
Carbon disulfide	NA	NA	No data available	95.2

2. These VOC emission estimates are speciated using appropriate speciation profiles.

Please note that there are two types of definitions used for organic compound emissions from fugitive sources. These are

- VOCs, which encompass all organic compounds excluding methane (i.e., equivalent to non-methane organic compounds—NMVOCs).
- TOCs, which includes all organic compounds, including methane.

Under the National Pollutant Inventory (NPI), only emissions of VOCs are required to be reported. However, to calculate emissions of certain organic species, it may also be necessary to calculate emissions of TOC. It is important to consider these definitions when estimating equipment losses using the following methodologies.

1.41 Total VOC Estimation

Before emission estimation techniques (EET) are discussed, it is important to note that some refineries estimate fugitive losses using various "in-house" techniques.

1.41.1 Total VOC Emission Estimation Methodologies

There are three main techniques that can be utilized to quantify total VOC emissions from process fugitives:

1. Correlation equations;
2. Leak/no-leak emission factors; and
3. Average emission factors.

These methods have been ranked in terms of the level of information required by a facility. Correlation equations require the most data, while average emission factors require the least.

1.41.2 Necessary Data for Estimating VOC Emissions

Before any of the EET presented in this section can be used for characterizing fugitive emissions, the following minimum information is required:

1. The number of each type of component (i.e., valves, flanges, etc.) in each process unit;
2. The service each component is in (i.e., gas, light liquid, heavy liquid); and
3. The time period each component is in that particular service (e.g., h/year).

Depending on the EET chosen, additional information may be required. This is discussed in greater detail in the relevant section.

The following definitions should be used when determining the type of "service" (i.e., gas/vapor, light liquid, and heavy liquid) a particular piece of equipment is in, so that the appropriate emission factors are used:

- *Gas/vapor*: the material is in a gaseous state under operating conditions;
- *Light liquid*: the material is in a liquid state in which the sum of the concentrations of individual constituents with a vapor pressure over 0.3 kPa at 20 °C is greater than (or equal to) 20 weight percent (wt %); and
- *Heavy liquid*: the material does not fall under the classifications for gas/vapor or light liquid service.

1.42 Correlation Equations

This method can only be used if screening values (ppmv) are (or have been) obtained through a fugitive leak screening program.

The following points are important to note when using this methodology:

- emission estimates are for "total organic compounds" (TOC) and, as a consequence, a correction must be made to convert the estimates to VOCs (i.e., to exclude methane);
- these emission factors are on a "per source" basis; and
- each individual screening value must be entered into the correlation equation to predict emissions for an equipment piece. Do not average screening values and

then enter the average value into the correlation to estimate emissions. The following steps should be followed to determine fugitive emissions using the correlation equation approach.

Step 1: **Measure Leaks from Fugitive Sources**

For each piece of equipment tested, the recorded screening value will fall into one of the three categories. The correct estimation methodology must then be used for each category as follows:

1. For "zero" readings (i.e., no emission is detected), Step 2 should be consulted to estimate emissions;
2. For screening values between the lower and upper detection limits of the monitoring device, Step 3 should be used; and
3. For values greater than the upper detection limit of the monitoring device (i.e., a "pegged" emission reading), Step 4 should be used.

Step 2: **Use of Zero Default Factors**

If no emissions are detected by the FID (i.e., the measured level is below the lower detection limit), then the "default zero emission rate" emission factors are used, *unless* the lower detection limit of the monitoring device is greater than 1 ppmv.

In this case, half of the detection limit is used (See Table 1.35).

Step 3: **Use of Correlation Factors to Determine Leak**

If screening values (SV) are determined through testing (i.e., the measured level is between the lower and upper detection limits), then the "correlation equations" presented in Table 1.35 are to be used to determine the leak from each relevant component tested.

Step 4: **Use of Pegged Emission Rate**

If pegged screening values are detected (i.e., level is above the upper detection limit of the monitoring device), the "pegged emission rate" emission factors presented in Table 1.35 are used.

Step 5: **Correct TOC Readings to VOC Estimate**

Once emissions have been estimated from each source, the emissions must be converted from TOCs to VOCs. To do this, additional information is required on the approximate weight percent of VOCs and TOCs in the process streams from which the emissions originate. This is then combined with the emission estimate for each equipment component as follows:

$$\mathrm{EVOC} = \mathrm{ETOC}(\mathrm{WPVOC}/\mathrm{WPTOC}) \tag{1.59}$$

where

EVOC The VOC emission rate from the equipment (kg/h);

ETOC The TOC emission rate from the equipment (kg/h) calculated using the emission factors or correlations from Table 1.35.

WPVOC The concentration of VOC in the equipment in weight percent; and

WPTOC The concentration of TOC in the equipment in weight percent.

Table 1.35 Correlation equations for process fugitive emissions (these estimates are for total organic compound emissions and must therefore be scaled to exclude methane)

Equipment type	Default zero emission rate (kg/h)	Pegged emission rate		Correlation equation (kg/h) [SV is an abbreviation for screening value and is in units of ppmv (parts per million by volume)]
		10,000 ppmv	100,000 ppmv	
Connector (the category of “connector” pertains to non-flanged connectors)	7.5×10^{-6}	0.028	0.03	Leak $= 1.53 \times 10^{-6}\,(SV)^{0.735}$
Flange	3.1×10^{-7}	0.085	0.084	Leak $= 1.53 \times 10^{-6}\,(SV)^{0.703}$
Valve (note that the category of valves does not include pressure relief valves, as these are included under the category of “other”)	7.8×10^{-6}	0.064	0.14	Leak $= 2.29 \times 10^{-6}\,(SV)^{0.746}$
Open-ended line	2×10^{-6}	0.03	0.079	Leak $= 2.2 \times 10^{-6}\,(SV)^{0.704}$
Pump seal	2.4×10^{-5}	0.074	0.16	Leak $= 5.03 \times 10^{-5}\,(SV)^{0.61}$
Drain	1.5×10^{-3}	No data available	No data available	Leak $= 1.5 \times 10^{-4}\,(SV)^{1.02}$
Other [this “other” category should be applied to all equipment types that have not been specifically considered elsewhere in the table. These would include (but are not limited to) loading arms, pressure relief valves, stuffing boxes, vents, compressors, and dump lever arms]	4×10^{-6}	0.073	0.11	Leak $= 1.36 \times 10^{-5}\,(SV)^{0.589}$

If a number of equipment pieces can be grouped because they share the same process stream and thus have similar VOC/TOC ratios, TOC emissions can be added for this equipment group prior to performing Step 5, thereby helping to reduce the total number of calculations required.

Step 6: **Note Operational Hours**

For the specific equipment pieces tested, you should estimate the annual number of operational hours. This information is required to derive annual emissions based on the hourly emission rates.

Step 7: **Determine Total VOC Emissions**

If all process fugitive sources have been tested, total VOC emissions from all sources can be determined by adding the emissions from each individual equipment component. In some refineries, however, it may not be practical to screen all sources because of safety and/or cost considerations. Appendix A discusses the methodology used to estimate emissions from all components in a refinery when only a fraction of the source population has been screened.

1.43 Leak/No-Leak Method

As is the case with the correlation approach discussed above, screening using a portable monitoring device is required for this methodology. However, rather than recording discrete screening values, this approach relies on a “leak”/“no-leak” criteria. Note that the criteria used to define whether an equipment component is leaking can vary, although emission data are only available for a leak definition of 10,000 ppmv.

Step 1: **Measure Leaks from Fugitive Sources Using FID**

A leak is typically defined and recorded if a screening value of “greater” than 10,000 ppmv is returned by the monitoring instrument. Therefore, the emission factor chosen from Table 1.36 will depend on whether the component tested returns a *pass* (i.e., reading $>$ 10,000 ppmv), or *fail* (i.e., reading $<$ 10,000 ppmv).

Step 2: **Estimate the VOC Emission Rate**

Emissions are estimated for each of the equipment types listed in, using the following equation:

$$\mathrm{EVOC} = (\mathrm{FG} \times \mathrm{NG}) + (\mathrm{FL} \times \mathrm{NL}) \tag{1.60}$$

where

EVOC	The VOC emission rate for the equipment type (kg/h);
FG	The applicable emission factor for sources with screening values greater than, or equal to 10,000 ppmv (kg/hr/source);
NG	For the particular equipment type of concern, the number of sources with screening values greater than or equal to 10,000 ppmv;
FL	The applicable emission factor for sources with screening values less than 10,000 ppmv (kg/h/source);

Table 1.36 Leak/no-leak emission factors for process fugitives (these emission factors are for non-methane VOCs only)

Equipment type	Service	Leak (>10,000 ppmv) emission factor (kg/h)	No-leak (<10,000 ppmv) emission factor (kg/h)
Connectors (connectors are defined as *non-flanged* connectors.)/flanges	All	0.0375	0.00006
Valves (it is important to note that this equipment-type category does not include pressure relief valves)	Gas	0.2626	0.0006
	Light liquid	0.0852	0.0017
	Heavy liquid	0.00023	0.00023
Pump seals	Light liquid (the light liquid pump seal factor can be used to estimate leak rates from agitator seals)	0.437	0.012
	Heavy liquid	0.3885	0.0135
Compressors	Gas	1.608	0.0894
Pressure relief valves	Gas	1.691	0.0447
Open-ended lines	All	0.01195	0.0015

NL For the particular equipment type of concern, the number of sources with screening values less than 10,000 ppmv;

Step 3: **Note Operational Hours**

For the specific equipment pieces screened, the annual number of operational hours needs to be estimated. This is required to derive annual emissions based on the hourly emission rates.

Step 4: **Determine Total VOC Emissions**

If all process fugitive sources have been screened, total VOC emissions from all sources can be determined by adding emission rates from each individual equipment component. In some refineries, however, it may not be practical to screen all sources because of safety and/or cost considerations.

Table 1.36 presents the emission factors required to estimate emissions using the steps discussed above. You should note that "drains" are not included in this table as no emission factors could be found for such sources. If the drains are screened, and discrete data are recorded (in ppmv), the methodology should be used.

1.44 Average Emission Factors

Where no screening values are available for particular equipment types, the "average emission factors" presented in this subsection should be used. This methodology involves applying the following generic algorithm to estimate emissions from all sources in a stream, for a particular equipment type:

$$EVOC = FA \times WFVOC \times N \tag{1.61}$$

where

EVOC	Emission rate of VOC from all sources grouped in a particular equipment type and service (kg/h) (e.g., valves in light liquid service);
FA	Applicable average emission factor for the particular equipment type (from Table 1.37);
WFVOC	The average weight fraction of VOC in the stream;
N	The number of pieces of equipment grouped in the relevant category according to equipment type, service, and weight fraction of VOC.

Although the average emission factors are in terms of VOCs, the equation still requires an input regarding the weight fraction of VOC in the process stream (i.e., WFVOC) to account for any non-organic compounds. For example, if the stream contains water vapor, you will need to account for this in your calculations. An example calculation is presented at the end of this section, illustrating the application of this methodology.

Step 1: **Develop an Inventory of the Number and Service Type of Fugitive Sources**

The number and service type of each equipment type in the refinery must be determined. This is a prerequisite to the use of this methodology. To simplify data management, spreadsheets can be developed noting the types of equipment and service modes (i.e., gas, light liquid).

Step 2: **Group the Inventory into "Streams"**

To simplify calculations, it is recommended that the equipment/service mode combinations identified in Step 1 (e.g., valves in gas service) be grouped into "streams" according to the approximate weight fraction of VOCs (i.e., WFVOC) in each stream.

A further simplification (if possible) may be to group areas of the refinery according to the "average" weight fraction of VOCs in the process streams. It will be necessary to take account of the various service modes for each equipment type (e.g., gas, light liquid) contained within that area. Another approach may be to make the conservative assumption that all streams are approximately 100 % VOCs, thereby making WFVOC = 1.

Step 3: **Note Operational Hours**

For the specific equipment category defined by the above two steps, the number of operational hours needs to be estimated.

Table 1.37 Average emission factors for process fugitives (these emission factors are for non-methane VOCs only)

Equipment Type	Service	Emission factor (kg/h/source)
Non-flanged connectors	Gas	2.5×10^{-4}
	Light liquid	2.5×10^{-4}
	Heavy liquid	4.34×10^{-5}
Flanges	Gas	2.5×10^{-4}
	Light liquid	2.5×10^{-4}
	Heavy liquid	4.68×10^{-5}
Compressor seals	Gas	0.636
Pump seals	Light liquid (the light liquid pump seal factor can be used to estimate leak rates from agitator seals)	0.114
	Heavy liquid	3.49×10^{-3}
Valves	Gas	0.0268
	Light liquid	0.0109
	Heavy liquid	9.87×10^{-5}
Open-ended lines	All	2.3×10^{-3}
Pressure relief valves	Gas	0.16
Sampling connections	All	0.015
Drains	All	0.032
Other category (category should be applied to all equipment types in heavy liquid service that have not been specifically considered elsewhere in the table. Note, however, that some equipment types have emission factors applicable to "All" service types, which includes the heavy liquid category. Therefore, care must be taken to ensure these are NOT included under the "other" category)	Heavy liquid	5.18×10^{-5}

Step 4: **Use Emission Factors to Estimate Emission Rates**

Use the relevant emission factors and the equation given above to calculate the emissions from each equipment type. These emissions should then be added to derive a total emission rate for all equipment pieces quantified using this methodology. Table 1.37 presents the emission factors required to estimate emissions using the steps discussed above.

1.44.1 Example Calculation

The following example illustrates the application of the average emission factor approach:

1. A particular section of a refinery has a count of 300 valves (Step 1);
2. It is ascertained that 200 of these are in gas service (Step 1);
3. Within this smaller group of valves in gas service, it is ascertained that 100 valves are, on average, 80 weight percent VOCs, 10 % methane, and 10 % water vapor (Step 2);
4. It is estimated that this group of valves operates for 5,500 h per year (Step 3);
5. The appropriate emission factor for valves in gas service is 0.027 kg/hr/source (from Table 1.37) (Step 4). Emissions from this group of valves is thus estimated with the following parameters:
   (a) FA = 0.027;
   (b) WFVOC = 0.8 (as this equation parameter does not include methane or vapor); and
   (c) $N = 100$.
6. The final emission estimate for the group of 100 valves specified above is approximately 11,900 kg VOC/year.

The above steps would then be repeated again for the remaining 200 valves that were not included in the above estimate for that section of the refinery. In a similar manner, emissions need to be calculated from other potential fugitive emission sources in that section, followed by the next refinery "section," and so on until fugitive emissions from the entire refinery have been quantified.

1.45 Speciation of VOC Estimates

Once total VOC emission estimates have been determined, emissions can be speciated into NPI-listed substances according to either of the following two methods:

1. Use process stream composition data; and/or
2. Use limited speciation data in the form of weight fractions.

The first methodology is likely to give more accurate estimates than using generic weight fractions. In addition, the published speciation data are very limited, and therefore, a combination of these two methodologies may be required.

1.45.1 Speciation Based on Process Stream Composition

This methodology involves determining the organic composition of each process stream and applying these data to determine the vapor-phase composition.

This EET relies on the following equation to speciate emissions from a single equipment piece:

$$\mathrm{Ei} = \mathrm{EVOC} \times (\mathrm{WPi}\ /\mathrm{WPVOC}) \tag{1.62}$$

where

Ei	The mass emission rate of NPI substance "i" from the equipment (kg/h);
EVOC	The total VOC mass emission rate from the piece of equipment;
WPi	The concentration of NPI substance "i" in the equipment in weight percent;
WPVOC	The VOC concentration in the equipment in weight percent.

As for the techniques used to estimate total VOC emissions, it may be possible to group the refinery into sections according to process streams with similar compositions. "Average" composition data could then be derived from these individual process areas and used to speciate VOC emission estimates.

1.45.2 Speciation Using Developed Weight Fraction Data

As there is only a limited amount of published data on the speciation of VOC emissions, not all equipment types have been included in Table 1.38. This methodology can, therefore, only be used for the equipment types identified.

The speciation factors presented in Table 15 can be used to calculate emissions of NPI substances using the following equation:

$$\mathrm{Ei} = \mathrm{EVOC} \times (\mathrm{WPi}/100) \tag{1.63}$$

where

Ei	The mass emissions of NPI substance "i" from the equipment (kg/h);
EVOC	The total VOC mass emission rate from the piece of equipment;
WPi	The weight percent of the relevant compound in the vapor released from the equipment, as provided in Table 1.38.

Table 1.38 Speciation data for some substances from equipment fugitives

Refinery source	Compound	Weight percent in VOCs Released
Flanges	n-hexane	4.76
	Cyclohexane	0.14
	Xylenes	0.28
	Benzene	0.14
	Toluene	0.7
Valves	n-hexane	4.76
	Cyclohexane	0.14
	Xylenes	0.28
	Benzene	0.14
	Toluene	0.7
Pump seals	n-hexane	11.4
	Cyclohexane	0.52
	Xylenes	1.34
	Benzene	0.52
	Toluene	3.1
Compressor seals	n-hexane	5.3
Drains	n-hexane	12.2
	Benzene	2.47

1.46 Process Fugitive Controls

This section provides a general overview of the two principal techniques used in refineries to reduce emissions from process fugitives. These are as follows:

1. Modifying or replacing existing equipment (referred to as "equipment modification"); and
2. Implementing a leak detection and repair (LDAR) program.

The implementation of an LDAR program could lead to a reduction by 63 % of emissions from equipment leaks. The following describes the procedure to implement this program.

- **Identifying components**

Each regulated component must be assigned a unique identification number, recorded and located in the facility and on the piping and instrumentation diagrams.

- **Leak definition**

  Leak definition means the threshold standard (in ppm). It depends on regulation, component type, service, and monitoring interval. Leak definition can also be based on visual inspections and observations, sound, and smell. A leak is detected whenever the measured concentration (ppm) exceeds the leak definition.
- **Monitoring components**

  For many regulations with leak detection provisions, the method for monitoring to detect leaking components is EPA Reference Method 21. This procedure uses a portable detecting instrument. Monitoring intervals depend on component type and periodic leak rate but are typically weekly, monthly, quarterly, and yearly.
- **Repairing components**

  Components have to be repaired as soon as possible after the leak is detected. The following practices can be applied:
  - Tightening bonnet bolts.
  - Replacing bonnet bolts.
  - Tightening packing gland nuts.
  - Injecting lubricant into lubricated packing.

  If the repair of any component is technically infeasible without a process unit shutdown, the component may be placed on the delay of repair list.
- **Record keeping**

  For each regulated process, a list of ID number for all equipment subject, detailed schematics, equipment design specifications, piping and instrumentation diagrams and results of performance testing and leak detection monitoring must be maintained.

  For leaking equipment, records, instrument and operator ID numbers, and the date the leak was detected must be maintained. The dates of each repair attempt and an explanation of the attempted repair method are noted. The dates of successful repairs and results of monitoring tests to determine whether the repair was successful are included.

1.46.1 Equipment Modifications

There is a range of equipment modifications that can be used to reduce fugitive emissions. In general terms, these involve either installing additional equipment that eliminates or reduces emissions, or replacing existing equipment with sealless types. Table 1.39 provides approximate reduction efficiencies for common equipment modifications.

Note that these equipment modifications are not applicable to the "correlation equations" and "leak/no-leak method," as these techniques are based on equipment screening. However, if any equipment emissions are estimated using average emission factors, these reduction efficiencies can be applied as discussed in the following steps.

Table 1.39 Summary of equipment modifications

Equipment type	Modification	VOC emission reduction efficiency (%)
Pump seals	Seal-less design	100 (It is important to note that seal-less design equipment can be a significant source of emissions in the event of equipment failure. No methodology was found to estimate VOC releases in the event of such a failure)
	Closed-vent system	90 [The actual control efficiency of a closed-vent system is dependent on the percentage of VOC emissions collected, as well as the efficiency of the control device to which the vapors are routed (e.g., a flare or a combustion device, such as a furnace)]
	Dual mechanical seal with barrier fluid maintained at a higher pressure than the pumped fluid	100
Compressors	Closed-vent system	90 [The actual control efficiency of a closed-vent system is dependent on the percentage of VOC emissions collected, as well as the efficiency of the control device to which the vapors are routed (e.g., a flare or a combustion device)]
Connectors	Weld together	100
Valves	Seal-less design	100
Pressure relief devices	Closed-vent system	The control efficiency of closed-vent systems installed on a pressure relief device may be lower than closed-vent systems used on other sources (i.e., pumps and compressors)
	Rupture disk assembly	100
Sample connections	Closed-loop sampling	100
Open-ended lines	Blind, caps, plug, or second valve	100

Step 1: **Derive Total Fugitive VOC Emissions**

Total VOC emission estimates should be derived using the average emission factor methodology.

Step 2: **Subtract Relevant Equipment Pieces**

Identify those specific equipment pieces that utilize the modifications shown in Table 1.35. Estimate the uncontrolled emissions from these particular equipment pieces using the emission factors. Once emissions from these equipment pieces have been derived, subtract this VOC contribution from the total VOC estimate (derived from Step 1). This avoids any possible confusion with double counting of emission estimates.

Step 3: **Incorporate Controlled Equipment Fugitive Emissions**

Estimate controlled emission estimates from the equipment pieces identified in Step 2, using the reduction efficiencies from Table 16, and the following equation is given by

$$\textbf{Controlled Emission} = \textbf{Uncontrolled Emission}^{*}(\textbf{1} - -\textbf{Reduction Efficiency}/\textbf{100})$$

Once controlled emission estimates have been derived from these equipment pieces, add these controlled emission estimates to the total VOC emission estimate derived using Step 2.

Water Pollution Control

2

Abstract

When deleterious waste materials from the oil, gas, and chemical plants are discharged into streams or other bodies of water, they become pollutants. This chapter is intended to cover the safety and environmental control aspects as the minimum requirements for water pollution control in oil, gas, and chemical processing industries and production plants. The scope is accomplished under following titles: (1) Refinery Water Pollution Standard and Control; (2) Petrochemical Industry Water Pollution Sources, Standard and Control; (3) Organic Chemical Manufacturing; (4) Monitoring.

Keywords

Grease · Groundwater · Organic constituents · Oxygen demand · Pollution · Prevention · Turbidity · Sampling · Soil water · Spill · Wastewater · Water pollution

Wastewater from petroleum industry contains organic compounds, phenols, toxic metals, and other pollutants such as iron, dissolved and suspended solids, oil, cyanides, sulfides, and chlorine. In order to reduce these emissions, an accurate analysis of processes is necessary.

Pollutants or contaminants which enter a body of water can be divided into:

- *Degradable (non-conservative) pollutants*: Impurities which eventually decompose into harmless substances or which may be removed by treatment methods, that is, certain organic materials and chemicals, domestic sewage, heat, plant nutrients, most bacteria and viruses, certain sediments.
- *Non-degradable (conservative) pollutants*: Impurities which persist in the water environment and do not reduce in concentration unless diluted or removed through treatment, that is, certain organic and inorganic chemicals, salts, colloidal suspensions.

A. Bahadori, *Pollution Control in Oil, Gas and Chemical Plants*,

DOI: 10.1007/978-3-319-01234-6_2,

- *Hazardous waterborne pollutants*: Complex forms of deleterious wastes including toxic trace metals, certain inorganic and organic compounds.
- *Radionuclide pollutants*: Materials which have been subjected to a radioactive source.

2.1 Wastewater Characteristics and Classifications

Natural waters and wastewaters are characterized in terms of their physical, chemical, and biological composition. The principal physical properties and the chemical and biological constituents of wastewater and their sources are a lengthy list. Each designated water body should be controlled according to regulations which may be comprised of both basic and more detailed numerical criteria as briefly discussed below.

To the extent practical and possible, all bodies of water should attain the basic criteria of the "Five Freedoms from Pollution":

1. Free from suspended solids or other substances that enter the waters as a result of human activity and that will settle to form putrid or otherwise objectionable sludge deposits or that will adversely affect aquatic life.
2. Free from floating debris, oil, scum, and other floating materials entering the waters as a result of human activity in amounts sufficient to be unsightly or cause degradation.
3. Free from materials entering the waters as a result of human activity, producing color, odor, or other conditions in such degree as to create a nuisance.
4. Free from substances entering the waters as a result of human activity, in concentrations that are toxic or harmful to human, animal, or aquatic life and/or are rapidly lethal in the mixing zone.
5. Free from nutrients entering the waters as a result of human activity, in concentrations that create nuisance growths of aquatic weeds and algae.

The water contaminant parameters determined in refinery wastewaters include biochemical oxygen demand (BOD5), chemical oxygen demand (COD), oil, total suspended solids (TSS), ammonia (NH3), phenolics, hydrogen sulfide (H_2S), trace organics, and some heavy metals.

Table 2.1 shows the major sources of each of the contaminants. Process wastewaters contribute a portion of virtually all, while other sources have more specific contaminant discharges.

2.1.1 Water Free of Oil and Organic Material

This category includes boiler blowdown, effluent from cooling water and boiler feed water makeup units, rain water from oil-free areas, and cooling water which cannot have direct contact with oil.

Table 2.1 Wastewater pollutants and sources

PollutionS	Sources
Heavy metals	Process wastewater
	Tankage wastewater discharges, cooling tower blowdown (if chromate-type cooling water treatment chemicals are used)
NH_3, H_2S, trace organics	Process wastewater (particularly from fluid catalytic cracking unit and coker)
Total suspended solids	Process wastewater
	Cooling tower blowdown
	Ballast water tank flow drainage and runoff
BOD5, COD, Oil	Process wastewater, cooling tower blowdown (if hydrocarbons leak into cooling water system)
	Ballast water, tank flow drainage, and runoff
Phenolics	Process wastewater (particularly from fluid catalytic cracking unit)

2.1.2 Water Accidentally Contaminated with Oil

This category includes water streams that are normally free of oil, but may contain oil after an accident. These streams comprise rain water from tank farms, pipe alleys, and oil-free processing areas, once-through cooling water, etc.

2.1.3 Water Continuously Contaminated with Oil but with Soluble Organic Material

This category comprises rain water from oil processing areas, tank drain water, deballasting water, cooling water blowdown, flushing, and cleaning water.

2.1.4 Process Water

This water is in contact with process streams, originating from steam stripping, crude oil washing, and some chemical oil treatment processes. It contains variable amounts of oil and soluble material such as ammonium sulfide, phenols, thiophenols, organic acids, and inorganic salts such as sodium chloride.

2.1.5 Sanitary and Domestic Water

The end result of sanitation and domestic water used in the chemical and petroleum is termed as wastewater. Relevant to the above-mentioned facts, these five categories of water may need different treatments, and for this reason, water streams are often kept segregated in a modern refinery to reduce the cost of water treatment facilities. But in chemical and petroleum industries, different process could cause pollution which should be treated.

2.2 Water Pollution Terminals

Terminals are those storage facilities where refined petroleum products are received from either refineries or import facilities. Fuel is distributed from terminals by truck or rail to retailers or bulk users.

Terminals are the points where wholesalers, distributors, retailers, and other end users access petroleum products. All terminals have loading gantries and storage and can be supplied by pipeline, ship, and in some cases by road transport.

Import terminals, however, are only supplied by pipeline from refineries or ports. Community health and safety issues associated with the operation of terminal facilities may include potential public exposure to spills, fires, and explosions although the probability of large magnitude events directly associated with storage operations in well-designed and managed facilities is usually low. The likelihood of community exposure to chemical hazards may be greater during road, rail, or water transport activities associated with fuel delivery and distribution.

2.2.1 Wastewater Pollutant Sources Crude Oil Terminal

The onshore facilities for most crude terminals will consist of storage tanks and associated equipment for crude oil, ballast water, and sanitary water. Thus, the major environmental concern is contamination of wastewaters with oil and the treatment of the ballast and sanitary waters prior to discharge. The treating methods for oil-contaminated wastewaters include various types of gravity separators. It is most beneficial to segregate the dirty and clean waters and thereby to minimize the volume of water requiring treatment.

2.2.2 Product Terminal

Product terminals typically are separated from a refinery but in some cases may be associated with a refinery. The product typically handled at a terminal includes gasoline, diesel, fuel oil, liquefied petroleum gas (LPG), kerosene, aviation gasoline, and jet fuels.

The few environmental concerns encountered in a product terminal are similar to those in a refinery, and the pollution control methods for product terminals are similar to a refinery.

2.3 Design Procedure for Effluent Water Pollution Control

Examples of design and procedures which are generally beneficial are as follows:

- Recovery of oil spills and hydrocarbons with vacuum trucks to reduce emissions and water effluents.
- Separation of oily wastes, concentrated wastes, and other process wastes from general effluents for more effective treatment.
- Reduction of shock pollutant loads in treatment facilities through the periodic flushing of process sewers to prevent contaminant buildup and by the use of flow and load equalization prior to treatment.
- A specialized program for handling oily wastes, sludges, wash waters, and other effluents.
- Maximization of air fan cooling and employ cooling water only for those services in which low process temperatures make air fan cooling impractical or uneconomic.
- Limiting the amount of water used for process unit washdowns.
- Converting foul water strippers to reboiler strippers to reduce foul water and recover condensate.
- Using caustic injection into desalted crude to reduce NH_3 needed to control corrosion in the crude unit overhead system.

2.4 Spill Prevention and Control

Prevention of spills of oil and related petroleum products should be one of the prime objectives, both in the design and in the operation of the proposed facility, and should include, but not be limited to the sitting and design criteria for all facilities, operating procedures and their periodic review, inspection and monitoring of facilities, personnel training, revision of operating procedures (where required), and redesign of facilities (if necessary).

Among specific design parameters are impervious dikes around tankage (feedstock and product), containment of storm water from the process area(s), ability to treat contaminated storm water in the wastewater treatment facility, leak detection systems capable of detecting small volume or slow rates of leakage from the pipeline system and appropriate use of valves to minimize potential spill volumes.

2.4.1 Spill Prevention Techniques

Prevention of spills is the first line of defense in protecting life, property, and the environment. Experience has shown that operational or human error and equipment failures are the principal causes of spills. Both can be reduced through the involvement and commitment of all staff to spill prevention.

Proper design, inspection, and maintenance of general facilities are of principle importance. Operator capacity is also extremely important and must be periodically tested and upgraded.

Given good equipment, good operators, and good procedures, spills will be reduced. They will not, however, be eliminated.

The difference between a minor event and a catastrophic event depends almost entirely on planning. Such planning includes plant design with spill containment features, and alarms, a workable and efficient contingency plan, trained spill control personnel, and adequate spill control equipment.

2.4.2 Bulk Storage

Oil storage tank construction and material should be compatible with the oil stored and the storage conditions such as pressure, temperature, etc.

Impervious secondary containment should be provided for the capacity of the largest single tank plus a sufficient allowance for precipitation and free board.

New metallic tanks buried underground should be protected from corrosion by coatings, cathodic protection, or other effective methods compatible with local soil conditions. The use of non-metallic tanks, if available and practical, should be given consideration.

Above-ground tanks should be subjected to appropriate integrity testing. Appropriate procedures might include hydrostatic testing, visual inspections, or inspection by a system of non-destructive shell thickness gagging.

Plant effluents which are discharged into a watercourse should have disposal facilities observed frequently enough to detect any possible system upset that could cause an oil discharge.

2.4.3 Facility Drainage

Appropriate containment or diversionary structures should be provided to prevent oil from leaving the property uncontrolled.

2.5 Groundwater Pollution Control

Two basic sources of spilled liquid petroleum products are equipment failure and operator error.

- Equipment failure includes corrosion and leaking of both above-ground and below-ground piping and tanks, valves' failure, refinery unit upsets, and sewer and drain leaks. Many of these failures may be avoided through proper inspection and maintenance procedures.

- Operator error includes overfilling tanks and improper alignment of valves and piping. These and other operator errors can best be corrected through developing proven operating procedures, regular training and testing of personnel, and systematic follow-up to assure that procedures are followed.

2.5.1 Preventive Measures

The preventive measures to be installed during the construction of a permanent structure must consider the following:

- The type of construction (refinery, storage tank, pipeline, etc.).
- The volume and the nature of the oil likely to pollute the site.
- The geology and hydrogeological environment, nature of the terrain, depth, activity, and quality of the aquifer.
- The economic environment, proximity to and capacity of water wells and intakes for domestic purposes, risk of pollution of a river, etc.
- The preventive system involves four areas: corrosion protection, surface preventive measures, subsurface preventive measures, and monitoring devices to detect and warn of unsuspected pollution not visible from the surface or of a dangerous change in groundwater levels.

Other methods of preventing spilled product from entering the ground and controlling its direction are as follows:

- Rendering the soil impermeable where required by means of a concrete paving, a clay or bitumen layer, plastic sheets (PVC sheets covered with gravel, fiberglass reinforced epoxy), and chemical to be mixed with the soil.
- A surface drain system in the plant area carrying all oil and oil-contaminated water to a dirty water sewer and then to an interceptor or separator, by means of a pipe system with manholes (cast iron, steel, epoxy), and gutters.

2.5.2 Types of Devices

Trench

This system of protection, which is used as a barrier to prevent the horizontal movement of the oil, can only be carried out on a practical scale if the water table is situated at a depth of less than about 3–8 m depending on soil conditions.

Spread of oil on the groundwater surface is intercepted by digging the trench to about one meter below the piezometric level. Oil flows onto the water surface where it can be recovered.

Hydrodynamic protection

The principle of oil spill control using hydrodynamic methods is to effect a change in the groundwater flow pattern such that the free oil or the contaminated water, as the case may be, can be drawn to a specific control point or points. This can be achieved by discharging or recharging the aquifer, or a combination of both. The success of the method depends on maintaining an artificial gradient in the groundwater surface.

Monitoring

Groundwater monitoring devices are typically installed to detect and warn of unsuspected contamination not visible from the surface or of a dangerous change in groundwater levels. These devices are installed around petroleum storage areas, waste treatment/disposal facilities (including lagoons, land farms, and landfills), or an entire facility depending on the potential for contamination.

Care should be taken in choosing monitoring devices to maximize accuracy and reliability of the system. Also, monitoring should be conducted to differentiate between previous and new spills.

Mitigation measures

After a spill or any contamination is detected, remedial measures such as determination of the extent or contamination (boundaries) and a hydrogeological assessment of the contamination area to determine the corrective action necessary must be conducted. When an appropriate action is determined, steps including recovery of the oil and oily water and restoration of the site may be instituted.

Recovery measures

The principal factor to be considered when recovering free oil from the groundwater surface is to employ the natural water gradient or by inducing one or by increasing an existing gradient by artificial means. By good use and operation of recovery equipment, free oil can be concentrated at a relatively limited number of selected sites and removed. Wells and trenches are typically used for recovering oil and water.

2.6 Wastewater Pollution Control in Crude Oil Terminal

Wastewaters from petroleum industries are various. They can be process waters such as crude oil desalting waters or sour waters from hydrocracking or hydrotreatment processes, general effluents such as drained oily waters, washing waters, and finally spent caustics. In order to meet quality requirements about wastewater releases, the best way is to segregate these different waters. In this chapter, common techniques for wastewater treatments in refineries are shortly described, and then, best management practices for process wastewater are given.

The largest source of contaminated wastewater at a crude terminal is the ballast water from tankers. The quantity of ballast water requiring treatment depends upon the ship design, operation, and regulations governing the discharge of ballast waters. The ship design parameters include the amount of segregated ballast, the tank dimension, and the use of onboard oil/water separators. The operating parameters include the type of previous cargo, weather conditions, and tank cleaning. Optimizing the design and operation of a tanker can reduce the amount of water requiring treatment.

Ballast water treatment has consisted of settling the ballast in shoreside tankage for periods of 10–24 h, skimming off oil and discharging the water. This simple gravity separation may still be acceptable in some circumstances. For a better quality, effluent physical, chemical, and/or biological methods are necessary.

In some locations where shore space is at a premium offshore, deballasting facilities in the form of converted redundant tankers are utilized. One method to eliminate contaminated ballast water in tankers is the use of segregated ballast.

2.6.1 Simple Gravity Separation

Oil water separation is the first step of general treatment of residual refinery waters. Its purpose is to eliminate insoluble hydrocarbons and suspended matters. It is classically carried out by gravity. Several separators are available which can be longitudinal (API separators), circular, or lamellar.

These treatment systems rely on gravity difference to separate the oil and water. They are capable of removing the bulk of non-dissolved and non-emulsified oil. Examples include storage and settlement, once-through storage with skimming, API separator, corrugated plate interceptors (CPI), and holding basins.

2.6.2 Combination of Simple Gravity Separation Systems

Various combinations of the above individual treatment steps are used and such combinations may consist of storage and settlement plus API separators or CPI; storage and settlement plus holding basin; or storage and settlement plus API or CPI plus holding basin.

There are several reasons why a combination of steps is often the best choice of treatment process. First, storage and settlement ahead of a CPI or API will remove crude oil and prevent temporary overloading of the downstream separator.

Second, rather than sizing the CPI or API separator to handle the maximum hourly flow rate, costs can be reduced by\having a combination of a settling tank and CPI or API separator, with the tank designed for the maximum flow rate and the separator design to deal with the average flow rate. Third, a CPI or API plus holding basin serves as a guard chamber to finally trap any inadvertent discharge of oil from the settling tank.

2.6.3 Residual Suspended Matter

If it is necessary to reduce the non-dissolved oil content of the effluent below 25 mg/l, there are several processes available which can be added after the chosen simple gravity system.

Such methods will also reduce suspended solids to below approximately 30 mg/l. BOD can also be reduced because of the oil and solids removed. These processes have little effect on soluble oil content.

The physical methods include dissolved air flotation, filtration (using gravity or pressure filters), physical separation plus use of chemicals such as inorganic flocculants, and/or demulsifiers or polyelectrolytes, flocculation/sedimentation, flocculation-dissolved air flotation, and induced air flotation.

2.6.4 Physical and Chemical Purification

This step is necessary before biological treatment. This technique associates one chemical reaction with a physical separation. Most used techniques are coagulation, flocculation, air flotation, and filtration. It allows elimination of colloidal suspended matters and insoluble hydrocarbons.

2.6.5 Biological Treatment

After physical and chemical treatment, dissolved pollutants are still to be removed. These pollutants include soluble hydrocarbons, soluble CODs and BODs, phenols, and nitrogen compounds. They are biodegradable and can be removed with biological treatment techniques such as activated sludge or trickling filters. Biological treatment may in some cases be appropriate for removing dissolved biodegradable materials, which are often in low concentrations in normal ballast water. Typical devices used for biological treatment include activated sludge, trickling filters, rotating disks, and lagoons (aerated or not).

2.6.6 Spills

A major environmental concern with terminal operations is oil being spilled and the effects on birds and marine life.

Depending upon the type of terminal (offshore or onshore) and the characteristics of the water (such as currents and proximity to open water), the effects of a spill can vary from insignificant to extremely damaging. For example, an enclosed area such as an inlet from the sea, which has been described as the most productive marine environment, may experience accumulation of oil to unacceptable levels over time.

The oil that is spilled offshore will have less impact on the marine environment than an equivalent oil spill within an inlet from the sea (estuary) for three reasons:

1. There will be fewer organisms offshore to be affected.
2. The concentration of toxic compounds within the water column is expected to be less because more dilution water is available in offshore.

3. The contact time with the marine life will generally be shorter for an offshore spill due to the restricted flushing of the inlet from the sea.

In addition to spilled oil, the treated ballast water can also affect the marine life in an "enclosed" area. Thus, it may sometimes be best to pipe the treated water to another location where good mixing may occur, thereby protecting the "enclosed" area and minimizing the effects on the marine environment.

Spill contaminant is an important feature of a terminal. The most common types of barrier systems are used for floating booms and sorbent ropes. Two less frequently used alternatives are the air-bubble barrier systems and an enclosed berth.

2.7 Siting and Design

In previous sections, the sources of pollution and control methods were discussed. In the following pages, the general outline of pollutant specific for management in the siting and design of a refinery or terminal will be discussed.

A major design requirement for a refinery or terminal is to minimize or eliminate the emission of pollutants to the environment.

The methods necessary to implement this depend on the types of crude oil processed, the types of products, the availability of water fuel and other utilities, and the agreed pollution parameters.

Safety requirements must also be considered at this stage to ensure that the surrounding population and plant personnel are protected from hazards such as fire, explosion and toxic chemicals.

The items of concern in siting and design of a facility are briefly reviewed. They are meant only to point out potential environmental effects.

2.7.1 Aquatic Ecosystems

A refinery or terminal siting should take into account potential impacts to aquatic ecosystems. The characterization of aquatic systems in a siting should include: the location of spawning areas, feeding zones, commercial fishing areas, and sport fishing areas, a description of the benthic populations, and an estimate of primary productivity and its limiting factors.

Design considerations should include the facility's water supply requirements. While water use is usually non-consumptive, attention should be given to the intake and discharge zones.

At refineries or terminals, there is always the potential for a release of oil or petroleum products. All facilities should have a spill contingency plan and the basic pieces of equipment necessary to clean up a spill. All tankage should be properly diked.

2.7.2 Terrestrial Ecosystems

Impacts to terrestrial ecosystems from the siting of a refinery or terminal include a reduction or loss in total available habitat, destruction, or modification of food webs, and changes in populations. Another concern in the siting should be the sensitivity of plants and animals to pollutants.

2.7.3 Wetland Ecosystems

Where the water table is at, near, or above the land surface for a significant part of most years, the hydrologic regime is such that aquatic or hydrophytic vegetation is usually established, although tidal flats may be non-vegetated. Because wetlands are water systems, any alteration affecting the movement of quality of water in a small area is transmitted to other areas magnifying potential impacts.

Wetlands may be altered directly by filling, dredging, draining, or creating impoundments. Indirectly, alteration in water flow patterns at the locations and changes in adjacent land use can change the functions and values of wetland areas.

In addition to facility construction, the laying of pipelines associated with these facilities can have impacts on wetlands.

2.7.4 Land Use

Part of the siting for a refinery or terminal should consider the existing land use and compatibility with local and regional land use plans.

2.7.5 Water Pollution Control

The siting and design guidelines for water pollution control are as follows:

Siting

The most important aspect of water pollution control in the siting of a refinery is the effect of the wastewater effluent on the receiving water. Several factors which should be assessed in the siting investigation are as follows: heat load, total dissolved solids, heavy metal concentrations, and the effects of organics in the effluent.

Design

The design of the water pollution control depends on the degree of cleanup required to permit discharge of the wastewaters to either a body of water or public wastewater treatment plant and by the characteristics of the refinery. The following design practices are directed primarily toward segregation of process and non-process wastewaters and the recycling and reuse of raw and treated wastewaters:

(a) Refinery wastewaters should be segregated and treated based on oil content and potential for reuse. Four common divisions are oil-free wastewaters, oily cooling water, process water, and sanity wastewaters.

(b) Raw and treated wastewater streams should be recycled to reduce the effluent volume and thereby the makeup water required. Specific practices include the following:
use of catalytic cracker accumulator wastewaters rich in H_2S (sour waters) for making up to crude desalters after stripping;
(c) Tank farm, process, and product handling areas should be sloped and graded toward sumps or sewers to permit quick removal and collection of spills.
(d) Check valves and storage tanks should be installed at ends of pipelines specifically for spill prevention and control.
(e) Seals should be installed on normally closed valves.
(f) Cathodic protection systems should be installed for underground pipelines and tanks or apply a continuous coating around the pipeline to prevent direct contact between the pipe and the soil.
(g) Where feasible, all pipelines should be located above ground to facilitate inspection and identification of leaks.
(h) Concrete ditches, leading to either a chemical waste system or a wastewater system, should be installed under pipelines.

2.8 Source of Effluent in Petrochemical Industry

2.8.1 Water Pollution

In chemical plants, water formed by chemical reaction is generally less than the water evaporated into atmosphere, so that the water discharged tends to be less than the water intake. However, most part of the intake water is discharged.

Also, rain water is fouled while flowing through contaminated areas of a plant and is discharged as part of the wastewater.

2.8.2 Cooling Water

Most industrial water is used as cooling water. For example, there are the rapid cooling required for naphtha thermal cracking in making ethylene and EDC (ethylene dichloride) thermal cracking in the manufacture of vinyl chloride; removal of reaction heat in polymerization and oxidation; and cooling of fractionator condensers. Cooling water is not fouled since the indirect cooling method is generally adopted. However, there are instances of liquid leaking from the tubes of a cooler due to corrosion attack, resulting in contamination of cooling water.

2.8.3 Washing Water and Process Water

Considering the properties of fouled water, washing water is identical to process water, so the former is frequently called “process water.” Water is used in washing because it dissolves various substances. For example, carbon dioxide, hydrogen sulfide, and hydrochloric acid, in gas are dissolved in dilute alkali water.

The amounts of washing water and process water being used in petrochemical plants are small compared with those of cooling water. These wastewaters contain a considerable amount of organic substances and dissolved oils. But the COD and BOD values are insufficient to indicate the amounts of organic substance contained in the wastewater from the petrochemical processes.

It is necessary that the organic matter in wastewater from petrochemical plants be indicated using total organic carbon (TOC) and total oxygen demand (TOD), but the TOC and TOD indications themselves are not always considered useful as criteria for the detection of toxic substances in the wastewater or for the selection of wastewater treating methods. As organic compounds discharged from petrochemical processes have a variety of chemical properties depending on the processes, it is necessary to pursue the emission sources. Table 2.2 shows different types of pollutions in plants' operations.

Table 2.2 Types of water pollution

Pollution	Plant operation	Examples of pollution
Water pollution (wastewater, waste liquid)	Cooling water	Direct cooling or quenching of decomposed gas-wastewater contains tar dust, hydrogen sulfate and cyanide
		Breakage of indirect cooler tubes—contamination of cooling water due to liquid inside the tubes
		Steam ejector—steam condensate from ejector contains volatile hydrocarbon
	Boiler feed water	Steam ejector—steam condensate from ejector contains volatile hydrocarbon
	Washing water	Water containing hydrogen sulfate and hydrochloric acid, etc., is discharged from the washing of gas
		Water containing hydrochloric acid, etc., is discharged from the washing of liquid
		Dust is contained in discharged water from dust collector
	Process water	Solvent for suspension and emulsion polymerization Contain catalysts, emulsifier, plastic monomer, etc
	Feed water	
	(Chemical reaction, electrolysis)	
		Steam condensate from steam stripping contains dissolved hydrocarbons
		Steam condensate originating from dilution steam for naphtha thermal cracking contains carbon, phenol, and light oil
	Leakage (loss)	Leakage from pump and agitator shafts, valve stems, and flanges, and due to operational error

2.8.4 Typical Pollutants of Petrochemical Industry

Aqueous effluents originating from the petrochemical industry are essentially characterized by the presence of the following substances:

- Organic substances that are not biodegradable or only slightly so;
- Nitrogen compounds;
- Heavy metals.

2.8.5 Petrochemical Waste Treatment

Petrochemicals have been defined as bulk chemicals derived principally from natural gas, petroleum, or both. A careful check should be made of processes proposed or used for the manufacture of petrochemicals to decrease the possibility of water-soluble organics entering water supplies. The following methods should be considered:

- Recycling and reuse of waste streams;
- Quenching with oil or chemicals other than water that do not produce waterborne wastes;
- Use of alternative processes that do not produce waterborne wastes;
- Use of air coolers or of cooling towers in place of once-through cooling water;
- Elimination of waste products in the manufacturing operation before they become associated with waste streams;
- Processing of waste streams to reduce the amount of chemicals in wastewaters leaving the plant.

The extensive use of automated controls, alarms, and checks by the operators to prevent loss of chemicals is also important. It is essential that adequate facilities be installed to prevent uncontrolled release of chemicals and wastes to sewers or receiving waters. A very effective means of quality control is the use of large lagoons capable of holding several days' production of wastewater; this allows the water to be checked before being released to receiving waters.

2.8.6 Fertilizer

The liquid effluents arising from a fertilizer factory originate from a variety of sources and may be summarized as follows:

(a) Ammonia-bearing wastes from ammonia plants.
(b) Ammonia and urea wastes from urea manufacturing plant.
(c) Ammonium salts such as ammonium nitrate, ammonium sulfate, and ammonium phosphate.
(d) Phosphates and fluoride wastes from phosphate and superphosphate plants.
(e) Acidic spillages from sulfuric acid, nitric acid, and phosphoric acid plants.

(f) Spent solutions from the regeneration of ion exchange units, acid from cation exchange units and alkali from regeneration of anion exchange units.
(g) Phosphate, chromate, copper sulfate, and zinc wastes from cooling tower blowdown.
(h) Salt of metals such as iron, copper, manganese, molybdenum, and cobalt.
(i) Sludge discharged from clarifiers and backwash water from sand filters.
(j) Carbon slurry from partial oxidation units.
(k) Scrubber wastes from gas purification processes containing contaminants such as

Mono- and di-ethanol amines (MEA and DEA);
Arsenic as As_2O_3;
Potassium carbonate;
Caustic soda.

The effluents from fertilizer manufacturing are generated from a wide range of unit operations, and considerable variation between wastewaters from different factories may be noted. The age, state of repair, operational management, and degree of sophistication of each manufacturing unit will play an important role in determining the degree of in-plant materials loss, and the important factors leading to excessive losses (and subsequent pollution) may be summarized as follows:

- Outdated basic manufacturing plant with low efficiency and poor process control.
- Improper maintenance and repair, with particular emphasis on servicing of control equipment.
- Variations in feedstock and difficulties in adjusting process plant to cope with these variations effectively.
- Lack of consideration for pollution abatement and the prevention of material loss at the original plant design stage.

Overall water requirement for fertilizer manufacturing plants may be high, due to process cooling requirements. The total volume of effluent discharged is dependent to a considerable extent on the degree of in-plant recirculation, and in the case of total recycle, the raw water is used primarily for makeup purposes.

Plants designed on a once-through process cooling system generally give rise to high volumes of effluent, from 1,000 m^3 h-1 to volumes in excess of 10,000 m^3 h-1, consisting primarily of cooling water discharge.

2.8.7 Nitrogenous Fertilizers

A complex nitrogenous fertilizer plant based on the production of ammonium nitrate and urea products may give rise to pollutants such as ammonium nitrate, nitric acid, ammonia, urea, sulfuric acid, caustic soda, chromate, oil, grease, and boiler feed additives, contained within the overall effluent stream. The individual effluents from the ammonia and urea plants could be categorized as follows:

(a) **Ammonia plant**

HCN stripper outlet

Catalyst reduction
Shift process condensate

(b) **Urea plant**
Concentrate liquor
Cooling water blowdown
Additional pollutant discharges may arise from oily water and sanitary sewage effluents.

2.8.8 Phosphate Fertilizers

The examples of the sources of liquid effluents from phosphatic fertilizer manufacture are as follows:

Plant 1: Raw Water Intake

Effluent sources:
Superphosphate plant
Sulfuric acid plant
Water treatment plant
Cooling tower
Other sources

Plant 2: Raw Water Intake

Seawater (once-through cooling)

Effluent sources:
Ammonia plant
Phosphoric acid plant
Sulfuric acid plant
Utilities plant:
Cooling water makeup
Boiler blowdown, washings from ion exchange units and floor washing

2.8.9 Compound NPK (Nitrogen/Phosphorus/Potassium) Fertilizers

A major component of the NPK effluent source is the direct loss of fertilizer compounds from the granulation units. Examples of the effluents from compound fertilizer manufacture are as follows:

Plant 1: Raw Water Intake

Effluent sources:
Ammonia plant
Urea plant
Phosphoric acid and NPK plants
Water treatment plant
Cooling tower

Boiler plant

Plant 2: Raw Water Intake

Effluent sources:

Ammonia plant
Urea plant
Phosphoric acid and NPK plants
Sulfuric and nitric acid plants
Water treatment and steam generation plant
Methanol plant

2.8.10 Effect of Pollution

Wastes may be subcategorized into major and minor elements, but it should be noted that in specific instances, particular minor waste components may exercise significant polluting effects.

Major pollutants

General water pollution effects from fertilizer manufacturing wastes are dependent primarily on the elements, nitrogen and phosphorus, in their varying chemical forms.

2.8.11 Nitrogen

Compounds in following sections are related to nitrogen.

2.8.12 Ammoniacal Nitrogen and Urea

These two compounds are grouped together since urea may be hydrolyzed to ammoniacal nitrogen. The pollution problems which may be attributable to ammoniacal nitrogen include toxicity, oxygen demand, and eutrophication. Ammonia can be toxic to fish and other aquatic life forms at relatively low concentrations, and urea itself may be toxic to some aquatic life.

Ammoniacal nitrogen and urea may both be oxidized biologically. As such, their presence must be considered a potential oxygen demand in receiving water. In addition, ammoniacal nitrogen may act in its role as a fertilizer in an aquatic environment, leading to excess growth of algae and aquatic macrophytes and contributing toward accelerated eutrophication.

The presence of high ammonial levels may also cause problems if the receiving water is used for water supply purposes, due to chemical interference with chlorination (i.e., formation of chloramine intermediates) and resultant increase in chlorine demand.

2.8.13 Nitrate

The water pollution problems resulting from high nitrate levels may be categorized into eutrophication and public health effects. High levels of nitrate can give rise to increased eutrophication, leading to the promotion of growth of algae and macrophytes, adversely affecting water quality and amenity value. Health hazards related to nitrate in water used for supply purposes are considered to be infant methemoglobinemia and carcinogenic potential.

2.8.14 Phosphate

The presence of significant levels of phosphate has important effects on eutrophication. In terms of inorganic nutrient enrichment of receiving waters, phosphate may in many instances be more important than nitrogenous compounds, due to the fact that some forms of aquatic plant life may fix atmospheric nitrogen, so removing the absolute requirement for soluble forms of nitrogen to promote growth.

Under these circumstances, phosphate becomes the growth-limiting agent, and programmes to control eutrophication have generally sought to reduce available phosphate limits, to prevent excessive algal and macrophyte growth, with subsequent increase in nutrient retention.

2.8.15 Minor Constituents

In addition to pollution arising from the discharge of nitrogenous or phosphatic elements in liquid waste streams, pollution can be caused by a number of secondary waste components. The most important of these may be listed as follows:

Oil and grease
Hexavalent chromium
Arsenic
Fluoride

In specific instances, one or more of these individual pollutants may give rise to detrimental effects in receiving water, due primarily to toxicity, or can cause inhibition of nitrification. In addition, oil and grease may adversely affect the oxygen transfer characteristics of a watercourse.

2.8.16 Olefin Plants

2.8.16.1 Liquid Effluents

The main liquid effluent streams of olefin plants are oily drains, flare system pump-out, caustic oxidation effluent, dilution steam generator blowdown, and fouled condensate.

2.8.16.2 Oily Drains

The major oily drain sources are water draw-offs from vessels and hydroblast decoking wastewater.

Hydroblast wastewater producing during cleaning coke-fouled equipment contains entrained coke fines which are screened before discharge to the oily wastewater treatment system.

2.8.16.3 Flare System Effluent

Flare system is required for safe disposal of hydrocarbons which may be released at pressure during mal-operation and emergency conditions.

The hot flare knockout drum is primarily provided in order to receive liquids from the flare seal drum, excess fuel gas, hot relief, and hot blowdown.

2.8.16.4 Spent Caustic Neutralization

The cracked gas from the cracking furnaces is scrubbed in a caustic wash tower to remove in convenient remained gas including carbon dioxide, hydrogen sulfate, and mercaptans prior to further processing in the plant cold section.

The spent caustic is purged from the wash tower and is laden with sulfidic constituents plus volatile organic compounds (VOC) such as condensed oils and benzene. The spent caustic scrubbing liquor is commonly the most problematic waste stream generated by an olefin plant. This is due primarily to the sulfate concentrations which can range as high as 6 % (expressed as NaHS), depending on the cracking furnace feed stock and wash tower operation. The most effective means for on-side treatment of spent caustic is wet air oxidation which can achieve the oxidation of reactive sulfate to soluble thiosulfate, sulfate, and sulfate.

2.8.16.5 Fouled Condensate

Suspect condensate consists of all condensate streams from heat exchangers where hydrocarbon side pressure is more than the heating steam. A leakage in one of these heat exchangers will result in a hydrocarbon contamination of the condensate. Normally, the condensate flow to the surge tank is zero. The possibility constituents in suspect condensate are propane, propylene, butane, butadiene, and pentane with their concentrations related to the maximum solubility.

2.8.16.6 Dilution Steam Generator Blowdown

The main sources of wastewater in an olefin plant are as follows:

Process water below down during normal and upset operating conditions;
Process water from quench tower;
Surface runoff from possibly contaminated or non-contaminated areas.

The wastewater components are variable depending on respective sources. But the major components are oil, phenols, H_2S, and hydrocarbons.

2.8.17 Polymeric Plants

Polyethylene plant (HDPE/LLDPE/LDPE)

- **Process wastewater**

The process wastewater discharged from the bottom of the dryer scrubber entrained polymer fines typically has following characteristic:

Temperature <50 °C
BOD5 <100 ppm
COD <200 ppm
Suspended solid 100 ppm max.

- **Rainwater and floor washing water**

Rainwater and floor washing water shall be collected and treated separately according to their origin.

- **Polymerization Area**

The ceiling of polymerization area has been considered potentially polluted with polymer powder and oil that may be sometimes spilled from pumps, compressors, and other mechanical equipments.

The rainwater and the washing water falling on paved areas may carry away powder and oil and therefore shall be gathered to the basins separately, where the entrained material is retained.

- **Extrusion building area**

The floors of the extrusion buildings may be occasionally polluted by polymer/scrapes and lubricating oils.

Polypropylene plants (PP)

2.8.17.1 Process Wastewater Effluent

The sources of continuous process wastewater effluent are dryer scrubber and palletizing section. The wastewater contaminated by trace polymer fines leaves bottom of the dryer scrubber. It has typically the following characteristics:

pH, 6–8
Temperature 40–60 °C
Contaminant polymer powder.

The wastewater from palletizing section combined oil and trace suspended solid. It has typically the following characteristics:

PH 7–8
Suspended solids 25 ppm
Oil content 1 ppm max
COD 5–10 ppm
Temperature 50–80 °C.

2.8.18 Polyvinyl Chloride Plants

Polyvinyl chloride plants (PVC) is polymerized just as any other plastics. During the chemical process of the polymerization, an extremely large heat is generated. For removing heat, two emulsion and suspension-PVC (E&S-PVC) processes have been established. The wastewater effluent coming out of E–PVC plant is similar to S–PVC plant.

The sources of the wastewater effluent are polymerization reactor additive coating, vacuum station, and recovered vinyl chloride monomer. The wastewater streams coming out of these sources are collected in the wastewater settling basin which consists mainly of two compartments. The wastewater contaminated with wet PVC and VCM components have typically the following characteristics reported in Table 2.3:

2.8.19 Aromatic Plants

The effluent in aromatics plant mainly is coming out from naphtha hydrotreating and catalyst regeneration units in normal operating condition.

- **Naphtha hydrotreating unit**
  The main effluent is sour water coming from the reactor effluent separator drum. Depending on the naphtha quality as feed, sour water contains hydrocarbons, H_2S, and NH_3.
  The typical characteristic of sour water is specified as follows:
  Temperature 45 °C
  HC content 100 ppm wt
  H_2S content 50 ppm wt
  NH_3 20 ppm wt

In this case, the water collected and treated in batch stage by direct oxidizing with hydrogen peroxide in the sour water treatment unit.

- **Catalyst regeneration unit**
  Catalyst regeneration unit as a part of aromizing section regenerates continuously deactivated catalyst coming from aromizing unit and introduces to aromizing reactor automatically.

Table 2.3 Wastewater contaminated with wet PVC and VCM characteristics

Temperature	40–70 °C
pH	6–7
COD	150–250 ppm wt
BOD5	75–150 ppm wt
Wet PVC	15–20 mg per kg water
VCM	0.5–1 mg per kg water

The major effluent streams consist of water from washing drum, liquid from oxychlorination drum and liquid from dryer package.

(a) **Water from washing drum**

The major constituents of water from washing drum are carbon dioxide dissolved in water with tracing of carbonate sodium, hydroxide sodium, chloride sodium, oxychloride sodium.

(b) **Liquid from oxychlorination effluent drum**

The liquid effluent from oxychlorination drum consists of some sodium salts. The wholly specified composition of liquid are sodium combinations such sodium hydroxide, sodium carbonate, sodium oxychloride, and sodium chloride which shall be neutralized with hydrochloric acid.

2.9 Environmental Protection for Industrial Waste

All industrial complexes which produce waste in higher quantity above the standards should have waste treatment facilities before final release to environment. The dilution of treated wastewater to the standard level is not permitted. Monitoring system in final stage of treatment should be provided. Tables 2.4, 2.5, and 2.6 illustrate typical standards for municipal waste, effluent permissible concentrations, and characteristics of potable water.

2.10 Water Monitoring

Monitoring may be necessary for a number of reasons, such as

(a) To measure water problems (what, where, why, when).
(b) To measure waste parameters for use in calculation of waste treatment charges for municipal or regional treatment system.
(c) To measure waste parameters to allow detection and triggering of emergency actions in case of spills or process upsets.
(d) To measure the effect of a wastewater discharge on the quality of a receiving body of water.
(e) To measure the quantity and quality of all process wastes as well as influent liquid raw materials.
(f) To gather enough information on water used and contamination to allow design of pretreatment and/or reclamation systems.

Oil, gas, and chemical industries discharging to municipal or regional treatment systems shall be required to monitor their wastewater discharges or allow monitoring by others.

Table 2.4 Typical maximum effluent standard for municipal waste (daily average)

Effluent characteristic	Unit	Maximum concentration	Note
BOD5	mg/l	30	BOD5 should not increase to more than 50 mg/l
COD	mg/l	60	COD should not increase to more than 120 mg/l
Cl	mg/l	1	
Chloroform	MPN	100/100 ml	
Color	Color unit	16	
Detergents	mg/l	1.5	Equivalent to A.B.S
Dissolved oxygen	mg/l	2	
F	mg/l	2.5	
Ammonia (as N)	mg/l	2.5	
Nitrite (as N)	mg/l	50	
Nitrate (as N)	mg/l	10	
Oil and grease	mg/l	10	
pH	–	6.5–8.5	
Phosphate	mg/l	1	
Settleable solid	mg/l	0.1	
Suspended solid	mg/l	40	S.S should not increase to more than 60 mg/l
Sulfate	mg/l	400	Should not increase the sulfate content of supply water more than 10 %
Sulfite	mg/l	1	
Turbidity	N.T.U (nephelometric turbidity unit), (previously J.T.U.)	50	

2.10.1 Design Considerations for a Water Monitoring System

The following for questions when preparing to establish a water monitoring program should be answered:

(a) What must be or should be monitored?
(b) When should samples be collected and monitoring equipment located?
(c) When should samples be collected and monitoring equipment located?
(d) How should monitoring be done? (collection of samples, storage, analysis).

Table 2.5 A typical effluent permissible concentrations

CONTAMINANTS	Discharge to surface runoff, (mg/L)	Discharge to groundwater, (mg/L)	Irrigation and agriculture usage, (mg/L)
Al	5	5	5
Ba	2	1	1
Be	0.1	1	0.1
B	2	1	1
Cd	1	0.01	0.01
Ca	75	–	–
Cr^{6+}	1	1	1
Cr^{3+}	1	1	1
Co	1	1	0.05
Cu	1	1	0.2
Fe	3	0.5	5
Li	2.5	2.5	2.5
Mg	100	100	100
Mn	1	0.5	0.2
Hg	0	0	0
Mo	0.01	0.01	0.01
Ni	1	0.2	0.2
Pb	1	1	1
Se	1	0.01	0.02
Ag	1	0.05	0.01
Zn	2	2	2
Sn	2	2	–
V	0.1	0.1	0.1
AS	0.1	0.1	0.1
Cl^{-1}	Amount of chloride in industrial effluents should not exceed 250 mg/L (ppm) for freshwater	Amount of chloride in industrial effluents should not exceed 250 mg/L (ppm) for freshwater	Amount of chloride in industrial effluents should not exceed 250 mg/L (ppm) for freshwater
F	2.5	2	2
P	1	1	–
CN	0.2	0.02	0.02

(continued)

Table 2.5 (continued)

CONTAMINANTS	Discharge to surface runoff, (mg/L)	Discharge to groundwater, (mg/L)	Irrigation and agriculture usage, (mg/L)
C_5H_5OH	1	0	1
CH_2O	1	1	1
NH_4^+	2.5	0.5	–
NO_2^-	50	10	–
NO_3^-	50	1	–
SO_{42}^-	300	300	500
SO_3^{2-}	1	1	1
TSS	30	30	100
SS	0	0	0
TDS	Total dissolved solids in industrial effluents should not increase the amount of these materials more than 10 % in the underground water/river and any other sources in a distance of 200 m, in which effluent is dumped	Total dissolved solids in industrial effluents should not increase the amount of these materials more than 10 % in the underground water/river and any other sources in a distance of 200 m, in which effluent is dumped	Total dissolved solids in industrial effluents should not increase the amount of these materials more than 10 % in the underground water/river and any other sources in a distance of 200 m, in which effluent is dumped
Oil and grease	10	10	10
BOD	20	20	100
COD	50	50	200
DO	>2	>2	>2
ABS (detergent)	1.5	0.5	0.5
Turbidity	50	50	50
Color	Color of source water should not exceed more than 16 standard units due to industrial effluent, dumped	75 unit of color	75 unit of color
Temperature	Temperature of industrial effluent should not change the temperature of source water more than ±3 °C in a distance of 200 m	–	Temperature of industrial effluent should not change the temperature of source water more than ±3 °C in a distance of 200 m

(continued)

Table 2.5 (continued)

CONTAMINANTS	Discharge to surface runoff, (mg/L)	Discharge to groundwater, (mg/L)	Irrigation and agriculture usage, (mg/L)
pH	6.5–8.5	5–9	5–9
Radio actives	0	0	0
Digestible coliform	400/100 mL	400/100 mL	400/100 mL
MPN	1,000/100 mL	1,000/100 mL	1,000/100 mL

These four questions cannot always be completely answered in the sequence shown; usually, information is gathered and tabulated and several questions are answered simultaneously. Conflicts may arise with selection of a monitoring station at a hazardous location because the proper type of explosion proof equipment necessary for the area is not available or is prohibitively expensive. In this case, the "where" and "how" are not compatible and a compromise must be made. In general, the parameters and the equipment which should be monitored are covered in next sections of this chapter.

2.11 On-site Portable Instruments for Water Pollution Control

The sources of pollution and T.L.V (threshold limit value) of these pollutants in wastewater are described in previous sections. An effort has been made to present the most appropriate methods and equipment that apply as generally as possible for monitoring surface water, ground waters, cooling or circulating water, boiler water, boiler feed water, wastewater effluents after varying degrees of treatment, and untreated municipal or industrial wastewaters.

Here again, the effort has been made to present methods and equipment of the widest possible application, but when monitoring of sample of highly unusual composition is encountered, the equipment of this manual may require modification or may be wholly inappropriate.

All parameters which have been mentioned in previous sections of this chapter (Maximum Effluent Standard for Industrial Waste) could be monitored by this portable instrument excluding organic and radioactive substances.

This equipment has been developed to meet the need for a simplified, convenient, and accurate means of testing water in the field. The colorimetric tests are made with colorimeter using pre-calibrated meter scales for direct readout.

The volumetric tests are conducted by titration using a unique buret and titration stand with a precision screw plunger.

This device is shown at the right of the kit. It dispenses titration solution so that accurate, reliable results are obtained from test to test.

Table 2.6 A typical physical characteristic of potable water

Characteristic	Desired threshold limit value (T.L.V)	Maximum value
Color	5 units	5 units
Odor	2	3
Turbidity	5	25
pH	7–8.5	6.5–9.2
Chemical characteristic of potable water mg/L		
Characteristic	Desired threshold limit value (T.L.V)	Maximum value
As	0	0.05
Cd	0	0.01
Cn	0	0.05
Pb	0	0.1
Hg	0	0.001
Se	0	0.01
Cr	0	0.05
Ba	0	1
Ag	0	0.05
B	0	1
Hardness	150	500
Ca	75	200
Mg	50	150
Mn	0.05	05
Fe	0.3	1
Zn	5	15
Cr	0.5	1.5
SO_4	200	400
Cl	200	600
N	0.002	0.05
Detergent	0.1	0.2
P	0.1	0.2
Total dissolved solid	500	1,500

2.11.1 Alternative Current Colorimeter

As mentioned above, the basic of measurement is colorimetric which some reagent could be added to water sample to produce colored solution. This should be tested later. Type of portable instrument could be determined by relevant authorities.

See also the following Technical Data in Table 2.7 for more information:

2.11.2 Calibration and Inspection

- In general, this type of instrument is factory calibrated, but can be calibrated by competent person in accordance with the manufacturer's recommended standard.
- All reagent and chemical must be refreshed or in case of changing color and precipitation replaced by new one.

2.12 Online Fixed Measurement or Continuous Monitoring

The use of continuous monitoring equipment will permit the frequent measurement of several water quality parameters and the recording of this data at the measurement site or the transmitting of it to another location.

2.12.1 Continuous Water Sampling and Clarification System

The monitoring instruments should contain a sampling module, the sensors, a signal conditioning module, and a data logging or transmission module. A continuous flow of sample must be provided to the sensors by a submersible pump.

Each sensor in the instrument should be supplied with its own signal conditioner which converts the input to a standard electrical output. See Table 2.8 for more detail on sensors and ion-selective electrode.

A typical electrode placement for continuous monitoring is shown in Fig. 2.1. Further information has been given hereunder (see also ASTM Volume 11.01, 1989).

2.12.2 Calibration and Inspection

Calibration of the sensor signals should be carried out by adjustment of the controls on the signal conditioner.

The sample intake should be shielded with a screen and must be inspected according to manufacturer's instructions to prevent debris from clogging the system or damaging the pump.

Table 2.7 Technical data

Alkalinity	*Iron*
Titration for phenolphthalein and total alkalinity; enough of each indicator for 100 tests; enough titrant for 100 average (125 ppm) tests	Simplified phenanthroline method Colorimetric range: 0–3 ppm Enough reagent for approximately 100 tests
Carbon dioxide	*Manganese*
Standard titration procedure; enough reagent for approximately 100 tests	Cold periodate oxidation method Colorimetric range: 1–10 ppm Enough reagent for approximately 100 tests
Chloride	*Nitrate, nitrogen*
Mercuric nitrate titration—enough reagent for approximately colorimetric range: 0–1.5 ppm N, 100 tests; 0–15 ppm N, approximately 100 tests	Cadmium reduction–diazotization method; colorimetric range: 0–1.5 and 0–150 ppm N; enough reagent for 100 tests
Chlorine	*Nitrite, nitrogen*
Improved orthotolidine method; Colorimetric range: 0–1 ppm; enough reagent for 60 tests	Diazotization method; colorimetric range: 0–0.2 ppm N; enough regent for approximately 100 tests
Chromate	*Oxygen dissolved*
Diphenylcarbohydrazide method; improved Winkler method; alkaline-colorimetric range: 0–1.5 ppm; enough reagent for approximately 100 tests; Titration. 1 drop = 1 ppm	Improved Winkler method; alkaline-colorimetric range: 0–1.5 ppm iodide-azide modification; enough reagent for approximately 100 tests; Titration…1 drop = 1 ppm, tests. 1 drop = 0.2 ppm Enough reagent for approximately 100 tests.
Color	*pH, Wind range*
Colorimetric range: 0–500 APHA; platinum–cobalt units. No reagents required	Colorimetric range: 4.0–10; enough reagent for approximately 100 tests; no reagents required
Copper	*phosphate, ortho and meta*
Cuprethol method; colorimetric range: 0–3 ppm; enough reagent for approximately 100 tests	Stannous reduction method Colorimetric range: 0–3, 0–2, and 0–8 ppm; enough reagent for approximately 100 tests
Fluoride	*Silica*
SPADNS method; colorimetric range: 0–2 ppm Enough reagent for approximately 10 tests	Heteropoly blue method; colorimetric range: 0-3 ppm; enough reagent for approximately 100 tests

(continued)

Table 2.7 (continued)

Hardness, calcium	*Sulfate*
EDTA titration method; enough reagent for approximately 100 tests	Turbidimetric method; range: 0–300 ppm; enough reagent for approximately 100 tests.
Hardness, total	*Turbidity*
EDTA titration method 100 tests	Absorption method; range: 0–500 JTU; no reagents required
Hydrogen sulfide	
Screen test color chart comparison…0.1–5 ppm	
Enough test paper for 100 tests	

Table 2.8 Electrodes and ion-selective electrodes

Ion determined	Recommend Ref. electrode
Ammonia	Not applicable
Ammonium	Calomel
Barium	Calomel
Bromide	Double junction
Cadmium	Calomel
Calcium	Calomel
Chloride	Double junction
Copper	Calomel
Cyanide	Double junction
Fluoride	Double junction
Fluoroborate	Not applicable
Iodide	Double junction
Lithium	Double junction
Nitrate	Double junction
Oxygen	Not applicable
Potassium	Double junction
Silver/sulfide	Double junction
Sodium	Double junction
Sulfur	Not applicable

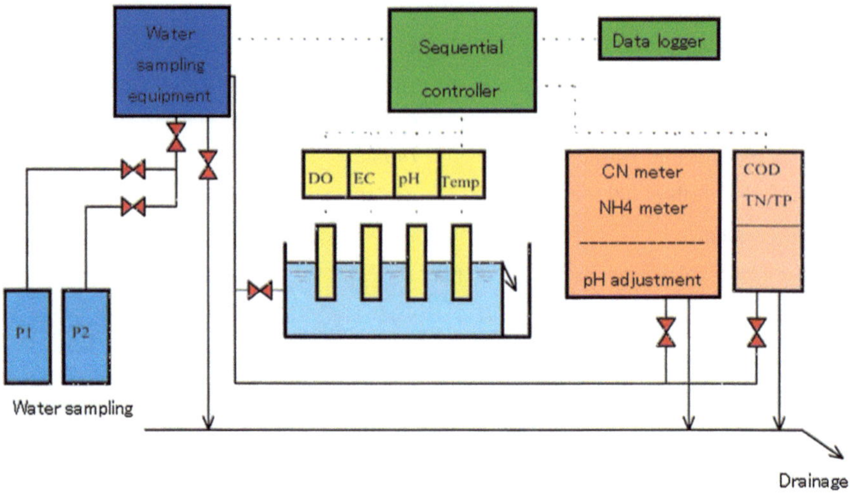

Fig. 2.1 Continuous water sampling and monitoring system

2.13 Laboratory Instruments

All parameters which are summarized in the following sections could be monitored in laboratory using instrumental or wet chemical methods.

2.13.1 Collection and Preservation of Samples

Sampling of different sources

The summary of different kinds of sampling is mentioned here, and for more details, see BS 6068-6.2, Water Quality Sampling.

Sampling of atmospheric precipitation

Atmospheric precipitation usually occurs in discrete events of limited duration.

Sampling precipitation equipment

Equipment for sampling precipitation includes a polyethylene polypropylene bucket and a stand.

Construction

The basic of this equipment is shown in Fig. 2.2. The equipment could be operated manually or automatically.

The equipment must be designed to minimize pollution during its operation. It should be manufactured from chemically inert materials to avoid contamination.

Surface water samples are usually taken directly into the sample container. It is not possible to take the sample by submerging the container by hand, laboratory forceps, or a holder with a sliding sleeve should be used.

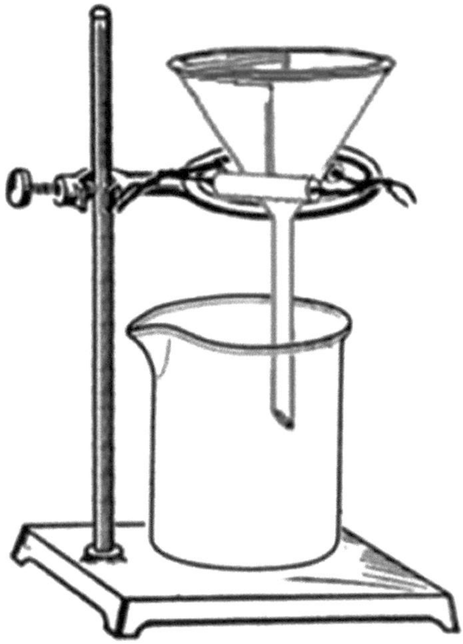

Fig. 2.2 Sampling of atmospheric precipitation

For such sampling, the container, which should have a volume of at least two liters, is usually made of stainless steel.

2.13.2 Soil Water Sampling

Soil water is defined as being all of the water contained in the soil.

Soil water sampling system

A common soil sampling system comprises a ceramic vessel with two pressure (vacuum) hoses inserted into it. The system operates by developing a vacuum through one hose, which causes the soil water to be drawn in through the porous walls of the vessel.

Construction

A schematic of soil water sampler is shown in Fig. 2.3.

2.13.3 Groundwater Sampling

The sampling of groundwater is usually carried out with devices and equipment very similar to that used for surface waters.

Similar containers are used for groundwater, mainly glass or polythene bottles.

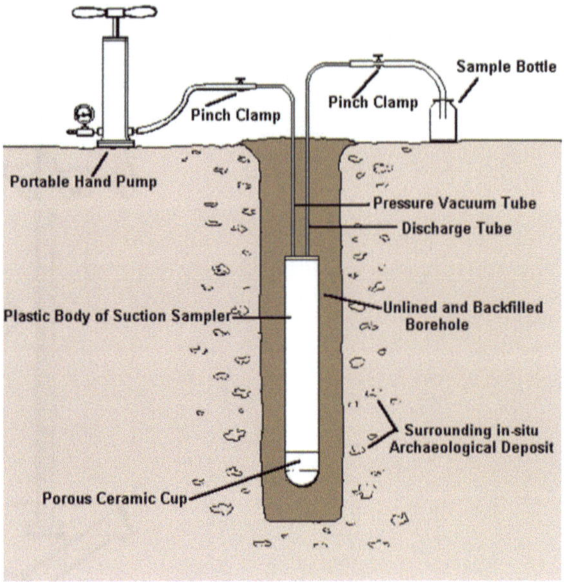

Fig. 2.3 The installation of a suction sampler

2.14 Physical Examination

2.14.1 Color

Color in water may result from the presence of metallic ion humus and peat materials, plankton, weed, and industrial waste.

The unit of color is produced by 1 mg/L platinum in the form of the chloroplatinate ion.

For color monitoring, see ASTM D 1882-00 and BS 2690: Part 9, 1970.

2.14.2 Conductivity

Conductivity is a numerical expression of the ability of a water sample to carry an electric current and is defined as μ moh/cm^2. This number depends on the total concentration of the ionized substances dissolved in the water. See also BS 2690: Part 9 1970.

Conductivity instruments

Measurement of electrical conductance is usually based on the use of conductivity bridges (Wheatstone bridge). The bridge circuit should always be arranged so that at the balance point, the detector indicates either zero or minimum potential. Basic of instrument is shown in Fig. 2.4.

In Fig. 2.4, we have

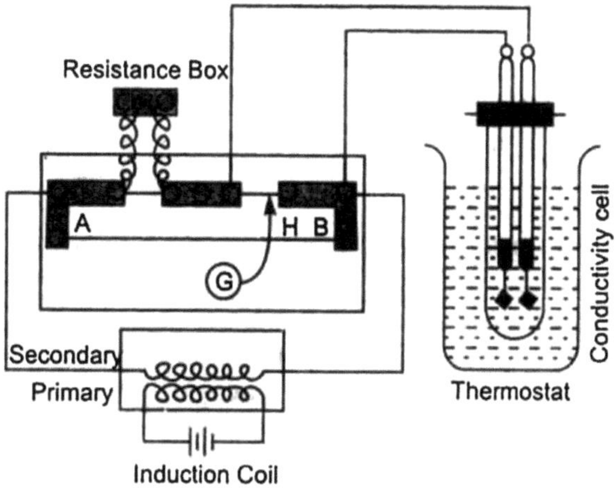

Fig. 2.4 "Inductive-type" conductivity cell

Determination of conductivity of the solution

We know that conductivity is the reciprocal of resistance and the resistance is more frequently determined by means of some form of Wheatstone bridge circuit. But, when direct current is passed through the solution, the following difficulties arise:

1. The concentration of the charges,
2. The polarization set in due to the substance liberated at electrodes. With the result, a back EMF is set up and the resistance of the electrolyte is altered. Kohlrausch (1868) avoided these difficulties as follows:

(a) Using an alternating current: The polarizing effects of the deposition of substances due to the current in one direction are neutralized by the effect of current in the other direction.

(b) Increasing the surface of electrodes: Polarization is decreased further if the surface of the electrodes is increased. This is done by coating electrodes in the cell with finely divided platinum.

(c) Using headphone: Since, with the passage of an alternating current, the ordinary galvanometer in Wheatstone bridge cannot be employed to determine the null point, it is replaced by a headphone.

(d) Using an oscillator: To obtain still better results, the induction coil is replaced by an oscillator having a frequency of 2,000–4,000 cycles per second.

Method

The solution of an electrolyte whose conductivity to be determined is taken as a special type of cell known as conductivity cell. This is made of Pyrex glass and is fitted with platinum electrodes. The electrodes usually consist of stout platinum plates sealed into glass tubes which pass through mercury placed in the tubes. The relative positions of the electrodes are fixed by cementing the tubes to the ebonite cover. The electrodes are coated with finely divided platinum. To maintain definite temperature, it is placed in a thermostat. Copper wires are dipped in mercury placed in glass tube to make connections.

Determination of specific conductivity

The cell is connected to resistance box, *R* on one side and thin uniform wire AB of Meter Bridge on the other; secondary of induction coil is connected to the ends of the V bridge, while the primary is connected to a battery. The headphone, *G*, is connected to a sliding key, *P*, and the binding screw in between the cell and resistance box.

The sliding key, *P* is placed near the middle. When the circuit is complete, a buzzing sound is heard in the headphone. Plugs are taken out from the resistance box. The sliding key is moved along the wire until the sound in the headphone is reduced to a minimum. Thus, point *H* is recorded. The observed conductivity of solution is then calculated by applying the following formula:

Resistance of solution = (BH/AH) × (Resistance, R)

Or

1/(Obs. Conductivity) = (BH/AH) × (Resistance)

Thus, AH and BH are measured on graduated scale and *R* in ohms from resistance box.

2.15 Turbidity

Turbidity is an expression of the optical property that causes light to be scattered and absorbed rather than transmitted in straight lines through sample.

Turbidity in water is caused by the presence of suspended matter, such as clay, silt, finely divided organic and inorganic matter, plankton, and other microscopic organisms. See also BS 2690: Part 9, 1970.

2.16 Determination of Metals

The presence of metals in potable water, domestic wastewater, and industrial effluents is a matter of serious concern. Metals could be monitored by atomic absorption spectroscopy, polarography, ICP, and colorimetric methods.

2.16.1 Atomic Absorption Spectroscopy

In atomic absorption, the sample is atomized into a flame, producing atomic vapor of the elements in question. The atoms absorb the radiation from the light source. The amount of light absorbed is proportional to the amount of the element.

Atomic absorption instrument

Figure 2.5 shows schematic diagram of an instrument.

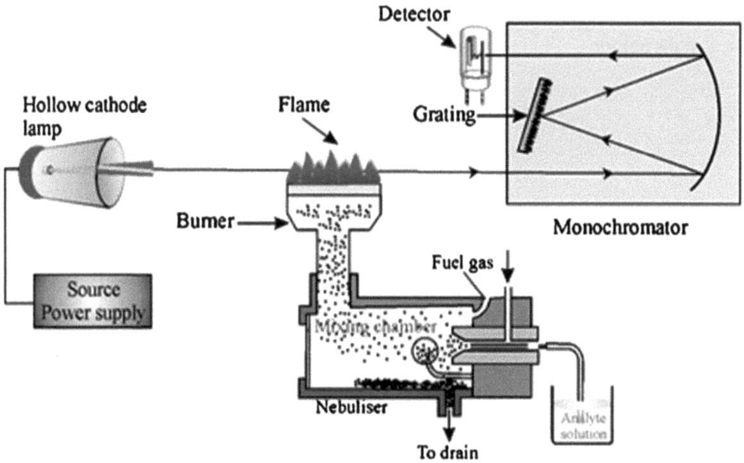

Fig. 2.5 Atomic absorption spectrometer

A mechanical modulated chopper alternately passes and reflects the light beam. One beam bypasses the sample and its intensity is measured as I0 and the sample's beam is measured as I1. The absorbed light is as I0–I1.

Calibration

The instrument should be calibrated according to manufacturer's manual book. The toxic pollutants such as Pb, As, Hg with low concentration (PPb) part per billion should be monitored by special device (vapor generation and hydride generation) which could be installed on atomic absorption instrument.

2.17 Polarography

Electrochemical methods are used for analysis of raw water and wastewaters.

2.17.1 Polarography Instrument

A schematic diagram of a classical polarography is shown in Fig. 2.6.

Polarography measurement is based on determining the time-averaged current of the dropping mercury electrode (DME) under diffusion condition.

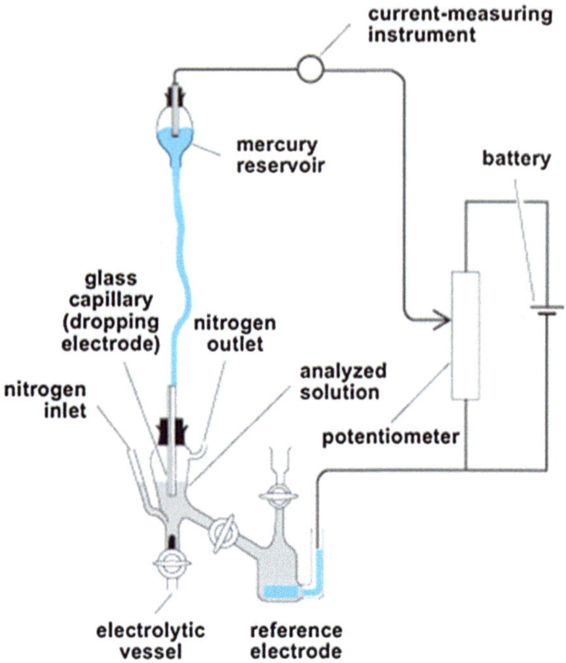

Fig. 2.6 Schematic diagram of classical polarograph

2.18 Chloride

Chloride, in the form of Cl ion, is one of the major inorganic anions in water and wastewater. Chloride could be monitored by titrimetric, potentiometric, and ion-selective electrode. More detail could be obtained from ASTM D 512-2004.

2.19 Chlorine (Residual)

The chlorination of water supplies and polluted water serves to destroy or deactivate disease, producing microorganisms. It could improve water quality by reaction with ammonia, iron, manganese, sulfide, and some organic substances. It should be monitored by titrimetric and colorimetric techniques. More detail should be obtained from ASTM D 1253-2003.

2.20 Cyanide, Fluoride, and Iodide

These three pollutants should be measured by titrimetric and/or colorimetric methods (UV–VIS spectroscopy) and ion-selective electrodes.

2.21 Nitrogen (Ammonia, Nitrate, Organic)

Nitrogen group should be monitored by titrimetric and/or colorimetric methods.

2.22 Ozone

Ozone should be measured by titrimetric method as given in ASTM Volume 11.02.

2.23 pH Value

The pH value of a solution is defined as the logarithm of the reciprocal of the hydrogen-ion concentration. Detail on effect of pH could be provided in API, manual on disposal of refinery wastes volume on liquid wastes, Chap. 2. pH should be monitored by electronic pH meter.

2.23.1 Calibration

It should be calibrated with standard buffer solution with known pH.

2.24 Phosphate

Phosphate content should be monitored by colorimetric method. For detail, see ASTM 11.02 and BS 6068 Sect. 2.28 1986.

2.25 Silica

Silica concentration could be measured by calorimetric and gravimetric method. For detail, see ASTM 11.02.

2.26 Sulfate

Sulfate could be measured by gravimetric method.

2.27 Sulfide

For sulfide monitoring, in general, it should be measured by colorimetric and titrimetric methods.

2.28 Determination of Organic Constituents

2.28.1 Grease and Oil

Solvent extraction infrared absorption

In this sort of monitoring, the oil is extracted from sample and the infrared absorption of the oil in solution is then measured.

Infrared spectrophotometer instrument

Radiation from the source is split into two beams: half passing into the sample cell and the other half into the reference cell. The reference beam then passes through the attenuator and onto the chopper. After dispersion by the prism or grating, the alternating beams fall on the detector and are converted to an electrical signal.

Schematic diagram is shown in Fig. 2.7.

Calibration of instrument should be according to manufacturer's manual book.

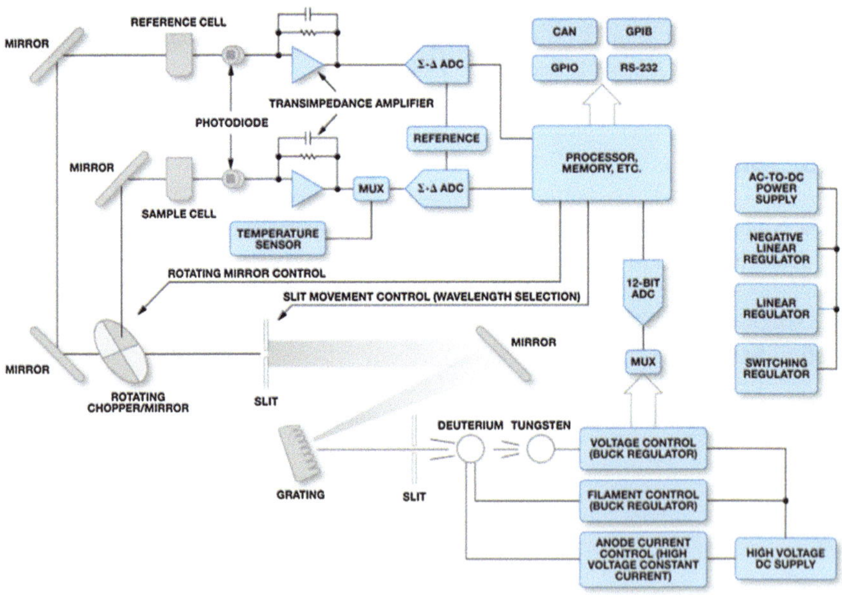

Fig. 2.7 Schematic diagram of a double-beam spectrophotometer

2.29 Combustible Gas Indicator

Equilibrium is established between methane in solution and the partial pressure of methane in the gas phase above the solution. The partial pressure of methane could be determined with a combustible gas indicator.

2.29.1 Combustible Gas Indicator Instrument

The heat generated by the oxidation of the gas increases the electrical resistance of the filament. The resulting imbalance of the electrical circuit causes deflection of a milliammeter.

2.30 Organic Carbon (Total)

Organic carbon TOC could be monitored by carbon analyzer in the range of 1–150 mg/L in water and wastewater.

2.30.1 Total Carbon Analyzer Instrument

Microportion of sample is injected into a heated packed tube in a stream of oxygen or purified air.

The water is vaporized and the organic matter is oxidized to carbon dioxide, which is measured by means of a non-dispersive type of infrared analyzer.

2.31 Oxygen Demand (Biochemical)

The biochemical oxygen demand (BOD) is a test in which standardized laboratory procedures should be used to determine the relative oxygen requirements of wastewaters, effluents, and polluted waters.

For this procedure, see ASTM Volume 11.02. See also BS 6068 Sect. 2.3 1984.

2.32 Oxygen Demand (Chemical)

COD is an important measured parameter for stream and industrial waste control.

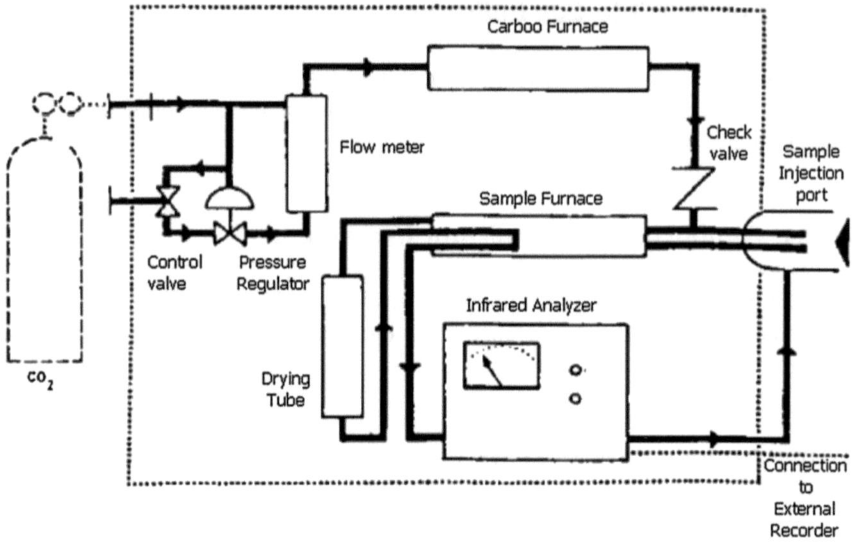

Fig. 2.8 Chemical oxygen demand (COD) analyzer flow diagram

2.32.1 COD Analyzer Instrument

Use dry CO_2 to carry the organic matter through a platinum catalytic combustion furnace which oxidizes it to CO and H_2O. Following the water removal and passage through a second catalytic treatment, the CO concentration is measured using an infrared analyzer. A schematic of instrument is shown in Fig. 2.8.

2.33 Examination of Water and Wastewater Radioactivity

The radioactivity in water and wastewater originates from natural and artificial or man-made sources. Artificial sources of radioactivity include fission, fusion, or particle acceleration, giving rise largely to alpha, beta, and gamma radioactivity.

2.33.1 Counting Room

The room should be free of dust and fumes that may affect the electrical stability of instrument. The background could be stabilized and lowered considerably by making the walls, floor, and ceiling out of several centimeters of concrete. Generally, temperature could be constant within 3 °C and should not exceed 30 °C.

Samples containing appreciable activity should be stored at a distance so as not to affect instrument background counting rate.

2.33.2 Alpha Particle Counter Instrument

Alpha particle counter consists of either a proportional detector or a scintillation detector, and a scaler conforming to the following requirements. For method, see ASTM 11.02 1989 and Technical Data.

Proportional detector

This may be one of the several types commercially available. The material used in the construction of the detector should be free from detectable radioactivity.

The manufacturer should supply voltage plateau and background counting rate data. Voltage plateau data should show the threshold voltage, slope, and length of plateau for a particular input sensitivity.

Scintillation detector

It should consist of an "activated" zinc sulfide phosphor having a minimum effective diameter of 36.5 mm. The phosphor should be mounted so that it could be attached and optically coupled to a multiplier phototube.

Scaler

Mechanical register, power supply, and amplifier are contained in a single chassis, generally termed the scaler.

Sample mounting disks or dishes

Having a flat bottom of a diameter slightly less than the inside diameter of the detector. Flat disks are preferred, but dishes could be used that have 3.2 mm high side walls. Platinum and stainless steel have been used for this purpose.

Calibration and standardization for general monitoring

Place a known amount of alpha standard into a volume of water sufficient to dissolve salts equivalent to those of the test samples and prepare for counting.

2.33.3 Beta-Particle Radioactivity Instrument

Beta-particle radioactivity of water and wastewater in general monitored by beta-particle counter which consists of the following components:

Detector

The end-window Geiger–Muller tube and the internal or external proportional gas-flow chambers are the two most commercially available types of detector.

Detector shield

The detector assembly shall be surrounded by an external radiation shield of massive metal equivalent to approximately 51 mm of lead and lined with 3.2 mm-thick aluminum.

2.33.4 Gamma-ray Monitoring

This section covers the monitoring equipment for gamma-ray-emitting radionuclides in water or wastewater by means of gamma-ray spectrometry.

Gamma-ray instrument

Gamma-ray spectra are measured with modular equipment consisting of a detector, and analyzer, memory, and a permanent data storage device. Lithium-drifted detectors, *p*-type or *n*-type, are used. A multichannel pulse-height analyzer should be used to determine the amplitude of each pulse originating in the detector.

2.34 Automated Laboratory Equipment for Monitoring Water and Wastewater

Automated instrument are available and in use to monitor individual samples at rate of 10–60 samples/hr. The same instruments could be modified to make analysis for multiple constituents simultaneously from one sample. The readout system includes sensing elements with indicators, alarms, and recorders.

Appropriate methodology could be supplied by the manufacturer for many of the common constituents of water and wastewater.

Table 2.9 Saturation (*S*) factors for calculating petroleum liquid loading losses

Cargo carrier	Mode of operation	*S* factor
Tank trucks and rail	Submerged loading of a clean cargo tank	0.5
Tank cars	Submerged loading: dedicated normal service	0.6
	Submerged loading: dedicated vapor balance service	1
	Splash loading of a clean cargo tank	1.45
	Splash loading: dedicated normal service	1.45
	Splash loading: dedicated vapor balance service	1
Marine vessels	Submerged loading: ships	0.2
	Submerged loading: barges	0.5

2.35 Loading Losses

2.35.1 Total VOC Estimation

Emissions from the loading petroleum liquid can be estimated (with a probable error of ±30 %) using the following equation:

$$LL = 0.12 \times SPM/T \quad (2.1)$$

where

LL VOC loading loss (kg/m^3 of liquid loaded);
S A saturation factor—see Table 2.9 below);
P True vapor pressure of liquid loaded (kilopascals (kPa));
M Molecular weight of vapors (kg/kg-mole);
T Temperature of bulk liquid loaded (K (i.e., °C + 273))

The saturation factor "S" accounts for the variations observed in emission rates from the different loading and unloading methods. Table 2.9 lists suggested saturation factors.

Emissions from controlled loading operations can be calculated by multiplying the uncontrolled emission rate (as determined using the equation above), by a reduction efficiency term:

$$\text{Controlled Emission} = \text{Uncontrolled Emission} \times (1 - \text{Efficiency}/100) \quad (2.2)$$

The overall reduction efficiency should account for the capture efficiency of the collection system, as well as both the efficiency and any downtime of the control device. These data should be obtained from the supplier or manufacturer of the collection system.

2.36 Emissions to Water

This section will be divided into two principal parts as follows:

- Point source wastewater discharges as released by refinery treatment plants; and
- Diffuse wastewater that arises through storm water and other miscellaneous runoff from the refinery site that is not captured and treated prior to discharge.

2.36.1 Point Source Discharge

The following tables should be used to provide "default" emission data for refinery effluent discharges that are not classified as transfers (transfers include discharging to sewer).

Table 2.10 Default speciation factors for organics in refinery effluent

Substance	Weight percent of DOC
Toluene	9.2×10^{-4}
Benzene	9.1×10^{-4}
Xylenes	1.4×10^{-3}
Phenol	6.9×10^{-4}
1,2-Dichloroethane	2.7×10^{-4}
Hexachlorobenzene	4.4×10^{-6}
PAHs	1.6×10^{-3}
Styrene	1×10^{-4}
Ethylbenzene	1.2×10^{-4}
1,1,2-trichloroethane	3.6×10^{-5}
Chloroform	2.5×10^{-3}

Table 2.11 Default emission factors for trace elements and inorganic in refinery effluent

Substance	Emission factors (kg/m^3 of flow)
Zinc	4.4×10^{-4}
Phosphorous	4.1×10^{-7}
Arsenic	6.7×10^{-6}
Chromium (VI)	7.7×10^{-6}
Selenium	3.1×10^{-6}
Nickel	3.6×10^{-6}
Copper	2.9×10^{-6}
Antimony	5.8×10^{-7}
Cobalt	1.6×10^{-6}
Mercury	1.1×10^{-8}
Cadmium	3.3×10^{-7}
Lead	1.9×10^{-6}
Cyanide	7.6×10^{-9}
Ammonia	1.3×10^{-6}

Based on discussions with the petroleum refining industry, the "dissolved organic carbon" (DOC) content of refinery effluents is a known parameter. Hence, the speciation factors for organic compounds in Table 2.10 are based on this parameter.

The document from which these data in Table 2.10 were derived indicates a ratio of DOC/COD of 0.267. In the absence of site-specific information regarding DOC, this ratio can be used to determine DOC from measurements of COD.

A similar parameter to DOC was not identified for trace elements and other inorganics in wastewater effluent. Therefore, trace elements and inorganic compound emissions are expressed as default emission factors in Table 2.11.

These speciation factors are applied to this effluent parameter in the following manner:

$$\mathrm{WWEi} = \mathrm{DOC} \times (\mathrm{WPi}/100) \times \mathrm{Flow} \tag{2.3}$$

where

WWEi	The wastewater emission of component "i" from the treatment plant (kg/hr);
DOC	The dissolved organic carbon (DOC) content of the treated effluent discharged by the plant (kg/m3);
WPi	The weight percent of component "i" as provided in Table 19 above;
Flow	The wastewater flowrate discharged to the receiving body of water (m^3/hr)

The emission factors in Table 2.11 are applied in the same way as factors are applied to air emissions, with the exception that they are based on the flow of effluent from the treatment plants (i.e., the emission factors are kg/m^3 of wastewater flow).

Soil Pollution Control

3

Abstract

When petroleum and chemical wastes are released on soil surface, hydrocarbons adsorb onto the organic mineral matter of the soil. Therefore, soil pollution with oil and chemical components has become an important problem of our days. The chemical and petroleum materials are released into soil environment by various ways such as pipeline blow-out. When the spill occurs, some kind of soil cleanup is typically necessary. The focus of this chapter is on long-term strategies for cleaning up petroleum and chemical products in unsaturated or saturated zones.

Keywords

Evaluation of leak · Excavation · Hydraulic · Leak detection · Saturated zone, site assessment · Soil flushing · Soil infiltration · Soil venting · Unsaturated zone

Cleaning up hydrocarbon-based liquids and refined products' release from pipelines, oil production units, oil desalter plants, pumping station, tank farms, chemical plants and drilling operation in upstream part and other pollutants from oil and petrochemical refineries and aboveground/underground storage tanks in downstream part of oil, gas and chemical processing industries typically involves using several corrective actions and strategies. Short-term emergency measures may involve imminent actions to control acute safety and health hazards such as potential explosions and toxications.

After the imminent danger has been eliminated, longer term corrective actions involve cleaning up pollutants that have entered the surface and subsurface environment.

Petroleum products in the subsurface may be trapped between soil particles in the unsaturated zone, floating on the water table or dissolved in ground water in the saturated zone.

A. Bahadori, *Pollution Control in Oil, Gas and Chemical Plants*,
DOI: 10.1007/978-3-319-01234-6_3,

In this chapter, the following matters of soil pollution control are discussed:

(a) To provide authorities concerned with information on how to assess site conditions in the unsaturated zone and where a petroleum product release has occurred, information needed to localize where in the unsaturated zone petroleum product is located, and also the removal of petroleum products from the unsaturated zone at a given site.
(b) To assess on the technologies designed specifically for cleanup of the saturated zone.
(c) To provide a structural methodology for evaluation and potential consequences of a leak in a pipeline. The methodology that is intended to assist pipeline operators in assessing the need to install pipeline leak detection facilities and an overview of available pipeline leak detection techniques.

3.1 Unsaturated Zone

Unsaturated zone is the portion of a porous medium, usually above the water table in an unconfined aquifer, within which the moisture content is less than saturation and the capillary pressure is less than atmospheric pressure and the unsaturated zone does not include the capillary fringe.

A good understanding of the conditions in the unsaturated zone is essential when selecting an appropriate soil treatment technology. By site assessment which reveals analysis of basic hydrologic, geologic, and chemical data measurements collected at the site of interest, the following points shall be considered.

(a) What was released? Where? (Time since petroleum released).
(b) Currently, where in the unsaturated zone is most of the petroleum product likely to be?
(c) How much petroleum product is likely to be present in different locations and phases?
(d) How mobile are the constituents of the contaminant, and where are they likely to travel and at what rate?

3.2 Site Assessment

The data collected should be used to gain the behavior of the contaminant in the subsurface. When dealing with the unsaturated zone, the data are primarily used to determine the mobility of the contaminant and which phase(s) it is likely to be in. Mobility of petroleum product in the unsaturated zone is related to the following:

(a) The potential for the various phases of contaminant to move through the subsurface; and
(b) The potential for the petroleum product to change from one phase to another.

To choose an effective cleanup technology for soils contaminated by petroleum, it is necessary to collect certain basic information about the released product and the subsurface environment and they are as follows:

1. **What contaminants were released?**
   Knowledge of the type of product released, its physical and chemical properties, and its major chemical constituents is required.

2. **Where is the petroleum product currently?**
   Distribution of petroleum product within the unsaturated zone, i.e., as a vapor in soil gas, as a residual liquid, or dissolved in pore water.

3. **How much petroleum product is in each phase?**
   Immediately after a release (weeks to months), most of the product will exist as a residual liquid and significant fractions will volatilize and dissolve in existing pore water or infiltrating rain water.

4. **Where is the petroleum product going?**
   Mobility effects not only extent the potential of a release, but also make the effectiveness of any treatment scheme that depends on mobilizing the contaminants, difficult to remove them (e.g. vacuum extraction).

These four questions provide a framework for decision, and more information can often improve the selection process.

3.3 Gathering Release Information

The questions listed in Table 3.1 provide a starting point for finding the mobility and phase distribution of the product in the subsurface. For more aboveground spills, the answers to questions in Table 3.1 are usually easy to obtain.

Petroleum products include a variety of fuel types, each with different physical and chemical properties. In addition, each fuel type is a mixture of many constituent compounds which have properties that can be quite different from those of the mixture. Mobility of different contaminants must be known to determine overspilled fuel (e.g., “Will this contaminant spread quickly and reach the water table?”), its partitioning (“Will this contaminant vaporize and pose explosion hazards?”), and its degradation potential (“Will this contaminant be biodegraded easily in the unsaturated zone?”). These factors are an important part of the technology selection process.

- **How much petroleum product was released?**
  The volume of spilled product can help the user to evaluate whether the contaminant has reached the saturated zone and estimate the level of contamination in the unsaturated zone.

Table 3.1 Basic release information assumed to be known

Information needed	Why information is important
What contaminants were released?	Physical and chemical properties differ for each contaminant, leading to varying phase partitioning, mobility, and degradation characteristics for each contaminant. Corrective action selection is tied to these characteristics
How much was released?	The amount released directly affects the phases in which in the contaminant may be found
What was the nature of the release (quick spill/slow leak)?	Phase partitioning and mobility of the released contaminant are both affected by the nature of the release. As a result, selection of appropriate corrective actions may differ for quick spills versus leaks over extended period of time
How long since the release?	Contaminants "weather" over time, that is, change in composition due to processes such as degradation, volatilization, and natural flushing from infiltrating rainfall. This change in composition directly affects the physical and chemical properties of the bulk contaminant
How was the release detected?	May provide insight into above questions and areal extent and distribution of contamination in the subsurface

- **Time since released**

  The time since release is important because the composition and properties of the released material change over time; volatile compounds evaporate, soluble constituents dissolve in infiltrating rainwater, and some constituents biodegrade. These biochemical changes that occur over time are called "weathering".
- **Gathering site-specific information**

  Site-specific information pertains primarily to the hydrologic and geologic characteristics of the site. Geologic characteristics can vary greatly, even over short distances; therefore, it is essential to make accurate estimate of soil parameters through numerous field data. Table 3.2 lists site-specific data that are needed to conduct a site assessment. This table lists the essential data needed to assess the site.

  Default values are obtained via tables, figures, and other data for many of the parameters. These default values enable users to make estimates of the critical parameters on timely basis for a preliminary site assessment.

  Site-specific data shall be obtained whenever possible. If a quick initial assessment is needed, it is often necessary to evaluate alternatives without the need of field data or when only incomplete data are available. In these cases, approximations are useful. Tables 3.3 and 3.4 provide typical values for several of the parameters listed in Table 3.2 for various types of soil and rock. These tables can be used to select default values in the absence of measured values. Figure 3.1 provides water-holding properties by soil type. An estimate of the moisture content

Table 3.2 Site-specific parameters to gather

Parameter (units)	Important for determining
Soil porosity (%)	Mobility, phase
Particle density (g/cm^3)	Mobility, phase
Bulk density (g/cm^3)	Mobility, phase
Hydraulic conductivity (cm/s)	Mobility, phase
Air conductivity	Mobility, phase
Permeability (cm^2)	Mobility, phase
Soil moisture content (%)	Mobility, phase
Local depth to groundwater (m)	Phase
Soil temperature (°C)	Mobility, phase
Soil pH	Bacterial activity
Rainfall, runoff, and infiltration rates (cm/day)	Mobility, phase composition
Soil surface area (m^2/g)	Mobility, phase
Organic content composition (%)	Mobility, phase
Fractures in rock	Mobility

Table 3.3 Factors to evaluate the mobility of liquid contaminants

Factor	Units	Increasing mobility (→)		
Release-related				
• Time since release	Months	Long (>12)	Medium (1–12)	Short (<1)
Site-related				
• Hydraulic conductivity	cm/sec	Low (<10^{-5})	Medium (10^{-5}–10^{-3})	High (>10^{-3})
• Soil porosity	% Soil volume	Low (<10)	Medium (10–30)	High (>30)
• Soil surface area	m^2/g	Low (<0.1)	Medium (0.1–1)	High (>1)
• Soil temperature	°C	Low (<10)	Medium (10–20)	High (>20)
• Rock fractures	–	Absent	–	Present
• Moisture content	% volume	Low (<10)	Medium (10–30)	High (>30)
Contaminant-related				
• Liquid viscosity	cPoise	Low (<2)	Medium (2–20)	High (>20)
• Liquid density	g/cm^3	Low (<1)	Medium (1, 2)	High (>2)

of a particular soil can be made using a typical field capacity value from the range of values shown in Fig. 3.2.

Table 3.4 Factors to evaluate the mobility of contaminant vapors

Factor	Units	Increasing mobility (→)		
Site-related				
Air-filled porosity (The total porosity minus fraction filled with water equals the air-filled porosity)	% volume	Low (<10)	Medium (10–30)	High (>30)
Total porosity	% volume	Low (<10)	Medium (10–30)	High (>30)
Water content	% volume	Low (<10)	Medium (10–30)	High (>30)
Depth blow surface	meters	Deep (>10)	Medium (2–10)	High (<2)
Contaminant-related				
Vapor density	g/m^3	Low (<50)	Medium (50–500)	High (>500)

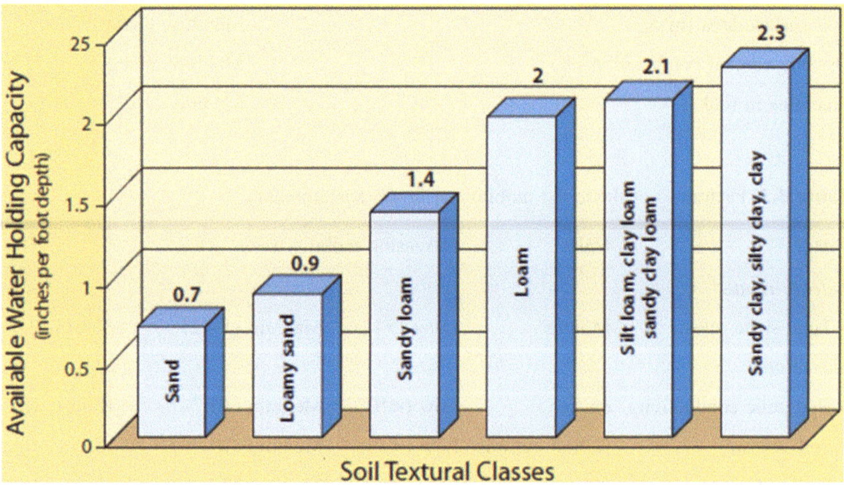

Fig. 3.1 Water-holding properties of various soils on the basis of their texture

3.4 Gathering Contaminant-Specific Information

In addition to release-related and site-related information, a site assessment shall include knowing of the physical and chemical properties of the contaminants released. To a large extent, contaminant properties govern partitioning in the subsurface; what phase(s) it is likely to reside in, how it is likely to move away from the site, and whether it is likely to degrade significantly over time.

Table 3.5 lists contaminant-specific data that are needed to conduct a site assessment. To illustrate, assume the liquid density of gasoline is needed. The user can locate liquid density in Table 3.5 and find the default source in Table 3.6.

Fig. 3.2 Scheme of a bioremediation system

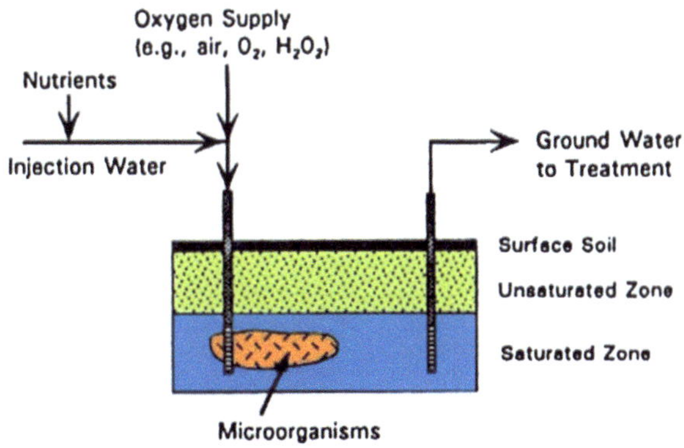

Moving to Table 3.6, the typical value for automotive gasoline in the liquid density column is 0.73 g/cm^3.

Hydraulic conductivity of soil directly affects a contaminant's mobility or ability to move away from the release site in the non-aqueous phase liquid (NAPL) and dissolved phases.

A high rainfall infiltration rate can cause contamination to move from one phase to another. Some hydrocarbons will dissolve in the infiltrating rain water, reducing the residual liquid portion of the contamination while increasing the amount of contaminant dissolved in pore water.

Soil temperature shall also affect contaminants' mobility, contaminants' vapor pressure, and therefore, the ease with which contaminants move into air spaces in soil increases with increasing temperature.

Table 3.5 Factors to evaluate the mobility of contaminants in pore water

Factor	Units	Increasing mobility (→)		
Site-related				
Hydraulic conductivity	cm/sec	Low (<10^{--})	Medium (10^{-5}–10^{-3})	High (>10^{-3})
Moisture content	% Volume	Low (<10)	Medium (10–30)	High (>30)
Rainfall infiltration rate	cm/day	Low (<0.05)	Medium (0.05–0.1)	High (>0.1)
Soil porosity	% Volume	Low (<10)	Medium (10–30)	High (>30)
Rock fractures	–	Absent	–	Present
Depth blow surface	meters	Shallow (<2)	Medium (2–10)	Deep (>10)
Contaminant-related				
Water solubility	mg/l	Low (<100)	Medium (100–1,000)	High (>1,000)

Table 3.6 Fluid properties

Fluid	Approx. density (kg/m^3) at 15 °C and 1 bar	Approx. vapor pressure [bar (abs)] at 0 °C
Crude oil (heavy)	875–1,000	0.3
Fuel oil	920–1,000	0.01
Gas oil/diesel	850	0.01
Crude oil (light)	700–875	0.55
Kerosene/naphtha/gasoline	700–790	0.01–1.2
NGL (condensate)	600–700	0.1–1
LPG	500–600	0.2–1.5
LNG	420	4–6
Ethylene	1.6	
Natural gas	1.1	
Sour natural gas (>0.5 % H_2S)	1.1	

3.5 Evaluating Contaminant Mobility

Mobility of the contaminants in a site assessment focuses on unsaturated zone and also between phases. Many in situ corrective actions depend on mobilizing contaminants.

The factors that control mobility of contaminants differ for each phase and are as given below:

(a) **Residual liquid contaminant**

The movement of bulk liquids in the unsaturated zone is dominated by three factors:

1. **Gravity**

It exerts a direct downward force, the magnitude of which depends only on the density of the contaminant.

2. **Pressure gradients**

They generally result from infiltrating liquid (precipitation and contaminants) and most often act in the same direction as gravity.

3. **Capillary suction**

It depends on the soil characteristics, and the forces it generates act in all direction, although not equally.

In addition to these three major forces, other physical, chemical, and environmental factors can influence a liquid's mobility in the unsaturated zone .

(b) **Contaminant vapors**

Vapors are generally mobile in the unsaturated zone. The degree of mobility greatly depends on the air-filled porosity of the soil. Several other factors also influence vapor transport in the unsaturated zone as listed.

Contaminant vapors may be mobilized by several natural or induced processes or forces as indicated below:

1. Bulk transport due to pressure gradients.
2. Bulk transport due to vapor density gradients.
3. In situ generation of gases or vapors.
4. Molecular diffusion due to concentration gradients.

Vapor density differences are important only when liquid contaminant of sufficient volatility is present, for example, the density of air saturated with gasoline vapors (i.e., in contact with liquid gasoline) is about 1,950 g/m^3 at 20 °C and moist air has a density of about 1,200 g/m^3 at 20 °C. In the absence of other driving forces, heavier (contaminant-containing) vapors should tend to migrate downward in the unsaturated zone. The driving force (density difference) will diminish as the vapors become more diluted.

(c) **Contaminant dissolved in pore water**

Table 3.5 lists factors that can help to determine the relative mobility of contaminants dissolved in pore water in the unsaturated zone.

As shown in Table 3.3, the site-related factors in Table 3.5 of this chapter may vary with depth. If information about the soil profile is available, this can be completed for each distinct soil group. Otherwise, a best estimate of actual conditions must be made to complete the preliminary assessment.

3.6 Technology Selection

Five technologies were used to determine UST sites as follows:

(a) Soil venting,
(b) Biorestoration,
(c) Soil flushing,
(d) Hydraulic barriers,
(e) Excavation.

3.6.1 Soil Venting

Soil venting is a general term that refers to any technique that removes contaminated vapors from the unsaturated zone. Venting may be passive or active. Passive venting is usually used for removal of methane gas.

Active venting uses an induced pressure gradient to move vapor through the soil and is more effective than passive venting.

3.6.2 Biorestoration

In situ biorestoration of the unsaturated zone is a process where oxygen and nutrients are added to contaminated soil to promote the breakdown of contaminants (see Fig. 3.2). In some cases, specially acclimated, commercially available bacteria may also be introduced to the subsurface, but this is not a common procedure. Bacteria capable of biodegrading petroleum hydrocarbons are commonly found in subsurface soils. Natural breakdown of petroleum hydrocarbons is likely to occur whenever they are introduced to the subsurface, but without the addition of nutrients and oxygen, biodegradation occurs very slowly.

3.6.3 Soil Flushing

Soil flushing refers to the in situ process where the zone of contamination is flooded with water or a water–surfactant mixture in order to dissolve the contaminants into the water or otherwise mobilize the residual contaminant to the water table.

The contaminants are then brought to the surface for treatment by strategically placed extraction wells. The wells must be located such that the groundwater is completely controlled hydraulically, to ensure that the leached or mobilized contaminants do not escape once they reach the groundwater. Figure 3.3 shows a schematic diagram of a soil flushing system.

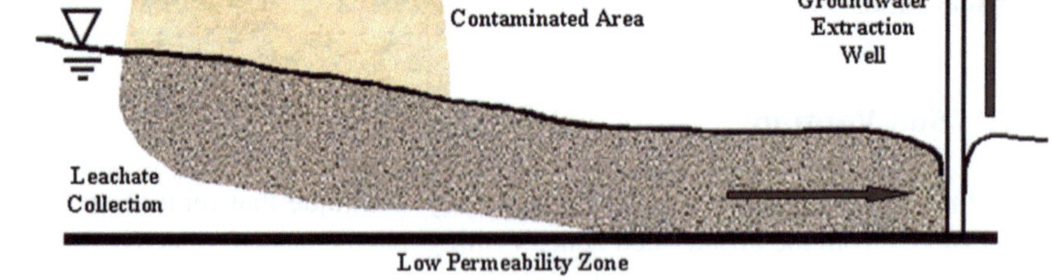

Fig. 3.3 Schematic diagram of soil flushing system

One soil flushing method removes contaminants using water to dissolve the liquid, sorbed, or vapor contaminants. These processes are influenced by the contaminants' solubilities and Henry's law constants.

A second soil flushing method mobilizes contaminants existing as free product in the soil pores and adsorbed to the soil. Contaminants found in these phases can be mobilized by the pressure gradient of the infiltrating flushing water.

The viscosity and density of contaminants control the extent to which a compound may be mobilized as free product. This type of soil flushing would be expected to remove a greater portion of gasoline than heating oil or #6 fuel oil, because gasoline is less viscous than either of the other two petroleum products. Many of the constituents of gasoline are also more soluble than those of #6 fuel oil or heating oil and would be more readily mobilized in the dissolved phase as well.

3.6.4 Hydraulic Methods

Typically, a trench is dug into the contaminated soil and residual liquid will began to seep into the trench. An impervious layer can be placed at the base of the trench to prevent re-infiltration of the product. As the product accumulates, it can be pumped out or removed manually, maintaining a gradient which facilitates further seepage into the trench.

3.6.5 Excavation

Excavation is an alternative to the in situ treatment methods. Excavation soil may be treated off-site, or disposed of (land field) without treatment. Treated soil is sometimes placed back in the excavation at the site. At present, excavating contaminated soil is more common than in situ treatment. Excavation, however, has many drawbacks that are not faced with in situ methods as follows:

(a) Excavating contaminated soils allows uncontrolled release of contaminant vapors to the atmosphere, increasing exposure risks;
(b) above and below ground structures, buried utility lines, sewers and water mains and buildings can pose real problems of contamination extends near or below the structures;
(c) aboveground treatment methods tend to be more expensive than in situ methods;
(d) disposal of contaminated soil is becoming increasingly difficult, and in some regulatory regions, the soil is considered a hazardous waste;
(e) a source of backfill is required to fill the excavation.

Excavation is a well-known technique, and it has the ability to remove most or all of the contamination from the site.

3.7 Saturated Zone

Saturated zone is defined as the zone below the ground surface in which all pore spaces are filled with water. Petroleum products reaching the saturated zone are subject to greatly increased mobility of the contaminant, particularly in the horizontal direction. Mobilization of the contaminants increases the magnitude of cleanup efforts and the opportunity for liability.

3.8 Site Assessment

After emergency measures have been accomplished at a site, the first step in a remediation plan shall be to gain a general view of conditions at the site. A preliminary investigation shall always be carried out before any sampling or analysis is specified. The aim of a preliminary investigation should be the following:

(a) To check the likelihood of contamination that may affect the suitability of the site for a specific use or any proposed future use;
(b) to identify whether any special procedures and precautions may need to be taken during operations;
(c) to provide the information from which an effective investigation of the site can be made.

3.8.1 Gathering Contaminant-Specific Information

Petroleum products are a mixture of many compounds. In general, remediation efforts are usually tied to one or more individual constituents that shall be targeted as follows:

(a) When the phase of contamination being evaluated is other than the original NAPL.
(b) If the mixture has "weathered considerably".
(c) Assessment on a compound that is thought to present the greatest potential threat.
(d) To design a treatment system.

9.4.2. Properties of the NAPL mixture shall be targeted as follows:

(a) When the phase of contamination being evaluated is the original NAPL;
(b) to evaluate the physical movement of NAPL through the subsurface.

3.8.2 Evaluating Contaminant Phase in the Saturated Zone

Petroleum products reaching the saturated zone will primarily be found in one of the three phases:

1. NAPL,
2. Dissolved in groundwater,
3. Sorbed to soil particles.

If the release is large, almost all of the petroleum product will exist as NAPL, but over a time, portion of the NAPL will transfer to the dissolved and sorbed phases.

If the release is small, all the NAPL may be trapped in the unsaturated zone.

- **Volume of floating NAPL**

Volume determinations of floating hydrocarbons are typically obtained by the following equation

$$V_N = A_N T_N n \tag{3.1}$$

where

AN is areal extent of the NAPL plume;
TN is average thickness of the NAPL plume; and
N is effective porosity of the soil formation.

NAPL thickness is fairly difficult to estimate accurately because the apparent thickness of the plume, as measured in a monitoring well, is typically greater than the true thickness. NAPL thickness will also vary throughout the plume, particularly if pumping wells are installed to contain the plume. Measuring plume thickness at several locations allows a better estimate of the average thickness.

Figure 3.4 is a schematic diagram that demonstrates the problem we face with the presence of LNAPL.

Figure 3.4 shows a monitoring well. The screen segment is perforated (or slotted) to allow entry to the groundwater. The horizontal, thick, black line says

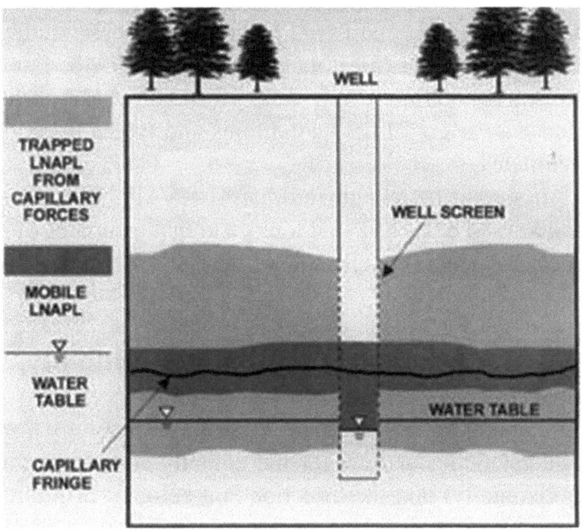

Fig. 3.4 Diagram to demonstrate the problem we face with the presence of LNAPL

"WATER TABLE": If there were no LNAPL in the area, the water face inside the well would coincide with this line.

However, the actual water face inside the well is DEPRESSED by the weight of the LNAPL that entered the well.

The next solid black lineup is crooked and says, "CAPILLARY FRINGE." If not for the LNAPL, this is the level in the soil to which the clear water would have risen.

The LNAPL resides on and presses down the water in the capillary fringe. The dark gray zone is where enough LNAPL accumulated so as to have mobility in the soil. The light gray above is where LNAPL resides in and tied to the soil by capillarity. In this zone, the LNAPL is not mobile.

The thickness of LNAPL in a monitoring well typically exceeds the thickness of the mobile LNAPL in the subsurface by a factor estimated to range between 2 and 10.

Due to this difference, the LNAPL thickness measured in a monitoring well is commonly referred to as the "apparent thickness" and is not an accurate measurement of the LNAPL thickness in the subsurface.

The monitoring well acts as a low point into which LNAPL drains. When LNAPL accumulates in the well, its weight depresses the water table in the well, resulting in additional LNAPL drainage into the well.

The difference between the actual and apparent LNAPL thicknesses increases with the increase in the capillary fringe, and the capillary fringe increases as the grain size of a formation gets smaller: the capillary fringe in silt being 1,000 mm, while in coarse grain sand, it is only 125 mm.

The result is that in a silty formation, for example, a 20-inch-thick layer of petroleum inside a well may represent a mere 2-inch layer of mobile petroleum in the formation, while in a coarse sand formation, these same 20 inches inside the well may represent 10 inches in the formation.

Many studies have been performed to correlate LNAPL thickness in a monitoring well to actual LNAPL thickness. These studies have produced correlations that can be used to estimate the actual LNAPL thickness from apparent LNAPL thickness measured in a well. However, these correlations may not be accurate under a variety of field conditions and typically produce only order-of-magnitude estimates.

It should be recognized that, if LNAPL is detected in a monitoring well, it is unlikely to be as bad as it looks and that estimates of release volumes derived from measurements in wells should take this LNAPL relationship into consideration.

3.9 Evaluating Contaminant Mobility

Implementing an effective remedial technology for saturated zone requires an understanding of how far the contaminants have travelled (vertically as well as horizontally) and the direction and velocity of plume movement.

3.9.1 Mass of Dissolved and Sorbed Contaminants in Groundwater

To estimate the mass of dissolved contaminants, the dissolved plume volume and average concentration must be known. This requires numerous samples from wells throughout the plume, both horizontally and vertically. Sorption of dissolved contaminants slows the center of mass of the plume relative to groundwater movement. This does not mean that the dissolved contaminant velocity is less than that of groundwater. Rather, the dissolved concentrations at the front of the plume are lessened due to adsorption, causing the plume's center of mass to move slower than groundwater flow. The liquid and dissolved contaminant phases are highly mobile, while sorbed contaminants are relatively immobile.

3.9.2 Extent of Contaminant Plume

Since the NAPL and dissolved contaminant can move at different rates and in different directions in the subsurface, the two phases may need to be delineated separately. Delineation of the downgradient and lateral extent of the plume is generally more important than of the upgradient extent because movement in the upgradient direction is generally limited.

Most petroleum products are less dense than water and will float on the water table in the bulk liquid phase. However, if denser than water, the contaminant will sink through the aquifer until it reaches an impermeable barrier. The areal extent of a denser-than-water contaminant plume is more difficult to estimate. Figure 3.5 shows a typical leaking underground tank with contaminant plume.

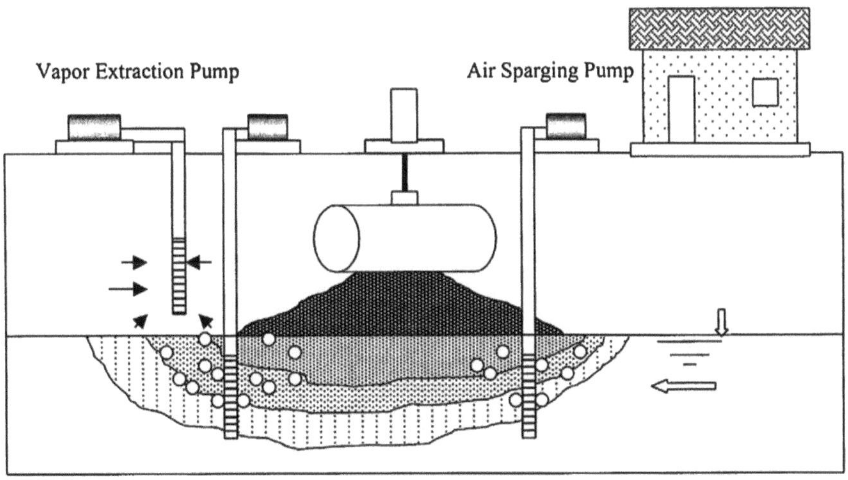

Fig. 3.5 A typical leaking underground tank with contaminant plume (Reprinted with permission © Elsevier, [90]

Floating product will move in the same general direction as groundwater, but movement of dense contaminants at the bottom of the aquifer is controlled by the grade of the impermeable barrier and may be different than groundwater movement. The flow direction of the denser-than-water bulk liquid is opposite to that of the groundwater, but the dissolved component travels with the groundwater.

If all the monitoring wells at a site are screened at or near the water table, they might not intercept a dissolved plume flowing deeper in the aquifer. Also, the entire dissolved plume may not be contained if the pumping rate is not adequate or recovery wells are not installed deep enough. A dissolved plume may exist relatively deep in the aquifer if the vertical component of groundwater flow is large. This is most prevalent at sites where the water table is relatively flat and the recharge rate is high. Infiltrating rain water acts to "push" the plume down below the water table.

3.9.3 Mobility of Contamination in the Saturated Zone

The important factors that influence NAPL and dissolved products to move through the subsurface are as follows:

(a) Subsurface stratigraphy, aquifer saturated thickness;
(b) local topography, location of nearby water bodies;
(c) location, depth, and pumping rates of nearby wells;
(d) regional and local groundwater flow direction(s), water table gradients (potentiometric head differences);
(e) hydraulic conductivity of formation(s); and
(f) density and viscosity of the bulk product or its constituents.

The above data can be used to estimate how quickly and in what directions contaminant plumes will travel. Groundwater flow direction is important in predicting where a contaminant is likely to migrate, and with seasonal changes, the water levels can affect flow patterns.

3.10 Setting Remediation Goals

An effective remediation plan should call for all phases of a cleanup. If the goal of remediation is to return the site to prerelease conditions; there are several typical cleanup phases that are likely to be implemented as follows:

(a) Emergency responses are a first priority to ensure that no immediate health or safety threat exists;
(b) containment of the dissolved and NAPL contaminants limits the extent of contamination and facilitates restoration efforts;
(c) removal and/or treatment of subsurface contamination must be accomplished; and

(d) finally, site monitoring program shall be set up to detect any changes in incomplete restoration.

3.11 Technology Selection

There are a variety of ways to direct effort to a release of petroleum product into the environment. No immediate action is necessary to a full-scale remediation effort to restore the environment to prerelease conditions. A typical remediation plan that has restoration to prerelease conditions have three main components as follows:

(a) **Containment of NAPL and/or dissolved product**

Migration of contaminants can be prevented through trench excavation which is effective only for containment of NAPL and also pumping wells which contain both NAPL and dissolved product.

(b) **NAPL removal**

NAPL recovery is often accomplished in conjunction with either trench excavation or pumping well installation.

(c) **Dissolved product removal**

Groundwater treatment methods can be either aboveground or in situ. The standard aboveground methods are air stripping and carbon adsorption. In situ treatment is where contaminants are treated or removed without disturbing the subsurface or bringing groundwater above ground.

3.12 Containing NAPL and/or Dissolved Contaminant

Two methods are described for preventing further migration of subsurface contaminants in this book:

(a) Trench excavation;
(b) pumping well installation.

3.12.1 Trench Excavation

A very simple way to prevent migration of floating NAPL is to dig a trench downgradient of the plume. As the NAPL reaches the trench, it can be intercepted, removed, and disposed of, thereby preventing migration beyond the trench.

At minimum, implementing a trench excavation containment system requires the following information:

(a) direction of plume;
(b) depth to water table;
(c) downgradient and lateral extent of the NAPL plume.

3.12.2 Pumping Well Installation

Pumping wells are an effective and commonly used method for containing a contaminant in the saturated zone. Unlike trenches, pumping wells are effective in containing dissolved contaminants and denser than water NAPL as well as floating NAPL. This method artificially lowers the water table at the site, drawing local groundwater and contaminants to the well.

The two most important considerations in this method are well location and pumping rate. To determine location and pumping rate, site hydrogeology and vertical and areal extent of contamination must be known. Each pumping well has a zone of contribution (ZOC), within which local groundwater will flow toward the well. To prevent contaminant migration, the well configuration and pumping rate must be such that the contaminant plume is completely contained within the well's ZOC. If the contaminant plume is very large or soil conditions prevent adequate pumping rates, several wells may be required to fully contain the plume. Hydraulic conductivity due to complex stratigraphy varies for different types of formations, and estimating the average hydraulic conductivity of such a site is difficult. Therefore, extent of the ZOC will not be equal in all directions and could result in poor location of wells.

3.12.3 Recovery of Floating NAPL

Recovery of NAPL floating on the water table is a common part of site remediation when a release of petroleum products reaches the saturated zone. There are several methods used to recover floating NAPL. The most common of these used are a pumping of oil/water separation scheme, depending on the containment method implemented at the site and vacuum extraction. Each of these methods is discussed below.

- **NAPL recovery with trench excavation**

  The two most common types of equipment used to recover NAPL from trenches are skimmers and filter separators. Skimmers float over the water surface and automatically pump NAPL off from water surface. Filter separators work much like skimmers except that the filter allows only petroleum product to pass through it.
- **NAPL recovery with pumping well installation**

  The two main types of recovery systems involving pumping wells are single-pump and dual-pump systems. In single-pump system, one pump is used to both contain the plume and recover the NAPL. In dual-pump systems, one pump is used to create a depression in the water table and another to remove NAPL floating on water table.
- **Vacuum extraction of floating NAPL**

  Vacuum extraction is commonly associated with treatment of the unsaturated zone. As the NAPL floats on the water table, some of the contaminant will transfer from the liquid phase to the vapor phase. The rate of natural volatilization depends primarily on the vapor pressure of the contaminants and the volume of air spaces in the soils above the NAPL plume. Vacuum extraction enhances natural volatilization by removing the vapors from the soil and bringing them to the surface. This creates disequilibrium between the liquid and vapor phases and allows volatilization to continue at a greater rate.

3.12.4 Treatment of Contaminants Dissolved in Groundwater

- **Aboveground treatment**

  Aboveground treatment of groundwater is generally accomplished by installing extraction or recovery wells that bring the groundwater to the surface where it can be treated (i.e., pump and treat). The two most commonly used aboveground treatment technologies are as follows:

(a) air stripping;
(b) carbon adsorption;

  which shall be discussed below.

- **Air stripping**

  Air stripping is the generic term used to describe several similar aboveground methods for treating contaminated groundwater. The facts behind this method are its cost-effectiveness, straightforward design, and also the concepts are well understood.

  There are four common air stripping methods as follows:

(a) **Diffused aeration**

  The contaminated groundwater is channeled through a large holding tank or pond with one or more diffuser pipes along the bottom. Air is pumped through the diffuser pipes and bubbles up through the tank, providing a medium into which the dissolved contaminant can volatilize.

(b) **Tray aeration**

  It is not as efficient in removing volatile organic as other air stripping methods, but because of its simplicity and low maintenance, it is sometimes used as a pretreatment for other methods.

(c) **Spray aeration**

  Contaminated groundwater is sprayed through nozzles over a pond or basin, greatly increasing the surface area of the water. As the water falls back to the pond, volatile organic compounds (VOCs) are transferred to the atmosphere.

(d) **Packed tower**

  Packed tower is one of the air stripping methods; pack tower works best at removing VOCs from groundwater, and they are often the most cost-effective method as well. Therefore, towers are the most widely used air stripping method.

  A typical packed tower is shown in Fig. 3.6. Contaminated groundwater is pumped to the top of the column and allowed to flow by gravity down through the packing material, while uncontaminated air is pumped up through the column.

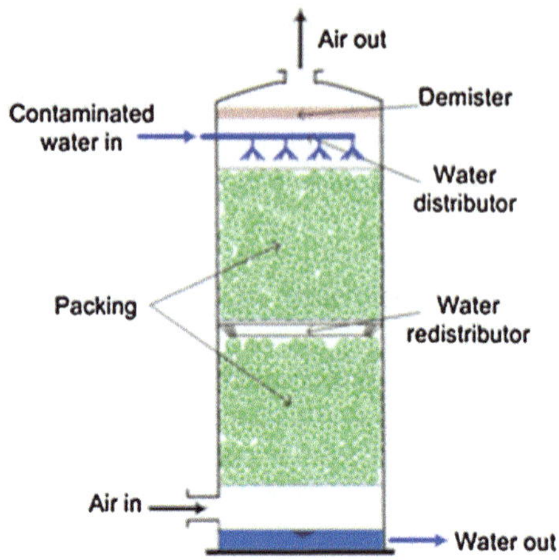

Fig. 3.6 Typical packed tower air stripper

- **Activated carbon adsorption**

Like air stripping, activated carbon is commonly used to remove VOCs from contaminated groundwater. The effectiveness of carbon as a treatment process is due to its ability to function as an adsorbent for molecules dissolved in water. This ability is largely due to carbon's large internal surface area.

Two forms of carbon are used to treat water:

(a) **Powdered Activated Carbon (PAC)**

It is dispersed freely into the water and subsequently filtered out. It used mostly as drinking water treatment but rarely for treatment of contaminated groundwater, because it is not easily reused.

(b) **Granular Activated Carbon (GAC)**

GAC is coarser than powdered carbon and is typically contained in a cylindrical unit. Contaminated water is passed through the unit until the adsorption capacity of the carbon is spent. The granular carbon can then be regenerated for future use.

Design of a GAC system is not as straightforward as for air stripping because of the complex interaction of the factors influencing removal efficiency.

3.13 Methodology for Pipeline Leak Consequence Evaluation

The methodology given in this book only evaluates the potential safety and environmental consequences of a leak and not the direct economic consequences related to repair and deferred production/transportation. The latter consequences are in general not reduced by the presence of a leak detection system and may be evaluated objectively, whereas the safety and environmental consequences are assessed, when necessary, on a more subjective basis.

The potential consequences of a leak are a function of various parameters related to the pipeline, its location, and the type of fluid that is transported. By comparing the outcome of the safety and environmental consequence evaluation with a leak consequence classification, the pipeline operator is able to determine the pipeline leak detection facilities required for the pipeline. The reduction in leak consequences as a result of a decreased time for leak detection and system shutdown can be demonstrated with the methodology given in this book.

The methodology presented in this book will not provide an absolute quantitative assessment of the consequences of a potential leak, but intends to rank pipelines on the basis of potential leak consequences.

A pipeline is not expected to leak if it is properly designed, constructed, operated, and maintained. Experience has shown, however, that despite all preventive measures taken, pipelines do occasionally leak. Therefore, even when authorities have no requirements for pipeline leak detection systems, the pipeline asset holder should formulate his own requirements based on a structured, quantified approach.

Proper pipeline management should ensure technical integrity of a pipeline in order to prevent failures and fluid releases. If a release does occur regardless of the measures taken to safeguard the pipeline integrity and the inspections to verify their effectiveness, e.g., by damage due to third-party activity, a leak detection system should make the operator aware of this.

A leak detection system in itself has no effect whatsoever on the leak expectancy (Le) of a pipeline and will only make the operator aware of the occurrence of a leak, enabling him to take remedial actions in order to limit the consequences of the release.

Installation of a leak detection system should not be given the first priority when a pipeline has a high leak expectancy; measures shall be taken to reduce the probability of a leak to as low as is reasonably practicable (ALARP principle).

The need for a leak detection system should not be related to the risk level itself but should be evaluated by assessing the potential reduction in failure risk due to reduced safety and environmental consequences in case a leak occurs.

3.13.1 Evaluation of Leak

The leak consequence evaluation in this book is a simplified version of the leak. The leak consequences are assessed by combining the following:

- factual input data, such as fluid pressure, density, etc.;
- assumed input data, such as most likely leak to hole size, time to detect a leak, and shutdown operation, etc.;
- factors such as fluid hazard factor, population density factor, etc.; and
- calculated parameters, such as fluid release rate, fluid release amount etc.,

The factors used in the assessment are based on expert operation.

The safety consequences are assessed based on the potential leak rate, the possibility of ignition, the population density, and the hazardous characteristics of the fluid and are expressed as a safety consequence factor (SCF). The environmental consequences are assessed on the basis of the potential leak volume, the persistence and/or seepage of the fluid into the environment, and the cleanup cost and other costs associated with the environmental consequences of a leak. The persistence and/or seepage of the fluid is adjusted by a climate correction factor. The environmental consequences are expressed as an environmental consequence factor (ECF).

Since the conditions will normally vary along the length of the pipeline, the pipeline is divided into sections and the safety and environmental consequences of a potential leak are evaluated for each section. For example, on offshore pipelines, the leak detection requirements are the highest close to the platform with regard to limiting safety consequences, while for environmental consequences, the requirements are often the highest in the shore approach area. Special leak detection techniques focusing on these different criteria should be used in the respective locations.

Other parameters will also vary along the pipeline, such as internal pressure, most likely hole size, time to detect a leak, water depth, etc. The worst conditions within a particular section are assumed to be valid over the whole section length.

The safety and environmental consequences of a leak are assumed to be pipeline section length dependent, since the potential number of leaks is length dependent.

The requirements for a leak detection system should be by classifying the safety and ECFs as "low", "medium" or "high".

The threshold levels between the "low" and "medium" and the "medium" and "high" categories have been initially defined but not verified and confirmed.

The need to install a leak detection system on pipelines is evaluated primarily on the basis of the safety and environmental consequences of a leak. The Le is applied in this evaluation as a secondary parameter, see Table 3.7.

3.13.2 Potential Leak Rate and Leak Mass

The actual amount of fluid released in case of a leak might range from very small to very large, depending on the leak rate, the presence of a leak detection system,

Table 3.7 Leak expectancy factor

Leak expectancy		Le factor
Input	Meaning	
HH	Very high	$\sqrt{3}$
H	High	$\sqrt{2}$
N	Neutral	1
L	Low	$1/\sqrt{2}$
LL	Very low	$1/\sqrt{3}$

the time to shut down pumps or compressors, and the presence and mode of operation of valves.

In the methodology, the leak size, i.e., the size of the hole in the pipeline wall, is a variable input parameter. The user can select the most likely hole size on the basis of potential failure modes, for example, hole of 50 mm caused by the impact of a tooth of a backhole excavator. The leak rate is then calculated as the fluid mass flow through the hole.

As part of the leak consequence evaluation, the potential leak mass is calculated using a number of assumptions. The calculated leak rate is assumed to continue until the leak is detected and the first remedial actions have been taken, such as closing block valves or shutting down pumps or compressors. The reducing leak rate during the remedial actions after valve closure or pump or compressor shut-down is not incorporated in this methodology, since this would complicate the assessment to a level which would be outside the scope of this leak consequence evaluation methodology.

(Note: The effect of this assumption is not as large as it seems, since for both gas and liquid lines, the consequences of a leak mainly refer to the time period between the onset of the leak and the system shutdown).

The safety consequences of a fluid release are governed by the leak rate, whereas the environmental consequences are related to the leak volume.

For onshore pipelines, the population density around the pipeline is evaluated on the basis of location classes as defined in ANSI/ASME B31.8:

Location Class 1: Areas such as wasteland, deserts, mountains, grazing land, farmland, and sparsely populated areas.

Location Class 2: Fringe areas around cities and towns, industrial areas, ranches, or country estates.

Location Class 3: Suburban housing developments, shopping centers, residential areas, industrial areas, and other populated areas not meeting Location Class 4 criteria.

Location Class 4: Areas where multistory buildings are prevalent and where traffic is heavy or dense and where there may be numerous other utilities underground.

For more detailed information about definition of these location classes, refer to ANSI/ASME B31.8.

For the safety consequences assessment for offshore pipelines, a distinction is drawn between the following pipeline locations:

- Open sea
- Shore approach
- Risers and pipeline sections on a platform and in the safety zone around a platform for an unmanned platform or complex.
- Similar to the above but for a manned platform or complex.

Table 3.8 gives a list of fluid types, with fluid hazard factors (S1) assigned to each fluid type. Table 3.9 lists the pipeline location classes and gives the

Table 3.8 Fluid hazard factors (S1)

Fluid	Approx. density (kg/m^3) at 15 °C and 1 bar	Approx. vapor pressure [bar (abs)] at 0 °C	Fluid hazard factor; S1
Crude oil (heavy)	875–1,000	0.3	0.5
Fuel oil	920–1,000	0.01	1
Gas oil/diesel	850	0.01	1
Crude oil (light)	700–875	0.55	1
Kerosene/naphtha/gasoline	700–790	0.01–1.2	5
NGL (condensate)	600–700	0.1–1	8
LPG	500–600	0.2–1.5	10
LNG	420	4–6	10
Ethylene	1.6		10
Natural gas	1.1		6
Sour natural gas (>0.5 % H_2S)	1.1		10

Table 3.9 Population density factor; S2

Area classification	Population density factor; S2
Onshore location class (as per ANSI/ASME B31.8)	
Class 1	1
Class 2	4
Class 3	8
Class 4	10
Offshore	
Open sea	1
Shore approach	5
Risers and safety zone (unmanned platforms)	6
Risers and safety zone (manned platforms)	10

population density factors (S2) assigned to the various locations. The SCF of a leak is calculated by:

$$SCF = L_R \times L_e \times l_g \times S_1 \times S_2 \times \left(\frac{L_s}{100}\right) \tag{3.2}$$

- Potential leak rate (LR)
- Ignition factor (Ig)
- Fluid hazard factor (S1)
- Population density factor (S2)
- Section length (Ls) in meters.

(Note: The value 100 merely reduces the result to a number between 1 and 1,000).

3.13.3 Environmental Consequence Factor

The environmental consequence factor (ECF) depends on the following parameters:

- Potential leak mass (Lm)
- Persistence/seepage factor (E1)
- Climate correction factor (E2)
- Cleanup and/or other associated costs (E3).

The ECF is only applicable to pipelines transporting liquids, i.e., air pollution and potential fire damage are not included in the evaluation.

The environmental consequences of a pipeline leak are expressed and quantified for practical reasons in terms of associated consequential money costs (USD) related to cleanup, compensation, etc. These depend on the volume of fluid released, the type of fluid, the environment category, and the type of contingency plans.

The amount of liquid released is adjusted by a factor which reflects the persistence of the liquid in the environment. This factor depends on the fluid type, the climate, and whether the leak is offshore or onshore.

It is assumed that the environmental damage is related to the liquid fluid components that remain in the environment after evaporation of the lighter fractions.

For onshore leaks, the damage to the environment caused by the persistence of the fluid is affected by the type of crude oil. Light crude will seep into the ground more easily than heavy crude, and it is therefore more harmful and more difficult and costly to remove.

The assessment of cleanup costs is based on costs for light crude. The cost for another fluid is calculated by multiplying the persistence factor for offshore leaks (which depend on the fluid density) or other persistence/seepage factor for onshore leaks.

It is assumed that in an offshore leak of light condensate with a density of 600 kg/m^3, or of any other hydrocarbon with a lower density, the total amount of fluid will evaporate under atmospheric conditions, and thus for those offshore leaks, the persistence factor is zero.

For denser fluids, the offshore persistence factor E1 may be calculated from the formula:

$$E_1 = 0.004 \times density\,(15\,^{\circ}\mathrm{C},\ 1\,bar) - 2.4 \tag{3.3}$$

The onshore persistence/seepage factor E1 may be calculated from the formula For density less than 850 kg/m^3:

$$E_1 = 0.0022 \times density\,(15\,^{\circ}\mathrm{C},\ 1\,bar) - 0.88 \tag{3.4}$$

For density greater than 850 kg/m^3:

$$E_1 = 0.0013 \times density\,(15\,^{\circ}\mathrm{C},\ 1\,bar) + 2.1 \tag{3.5}$$

The factors for onshore leaks are estimated from expert option. These relationships are shown in Fig. 3.7.

The climate correction factor, E2, in Table 3.10 allows for the relationship between liquid evaporation and the ambient temperature.

The cleanup cost and other associated costs for a liquid spill are expressed per unit volume in Table 3.11. The environment categories identified and associated cleanup cost and/or other consequential costs for light crude are given in US$/m^3.

The costs, quoted in 1993 money terms, have been estimated by experts using North Sea experience. They are assumed to be applicable worldwide.

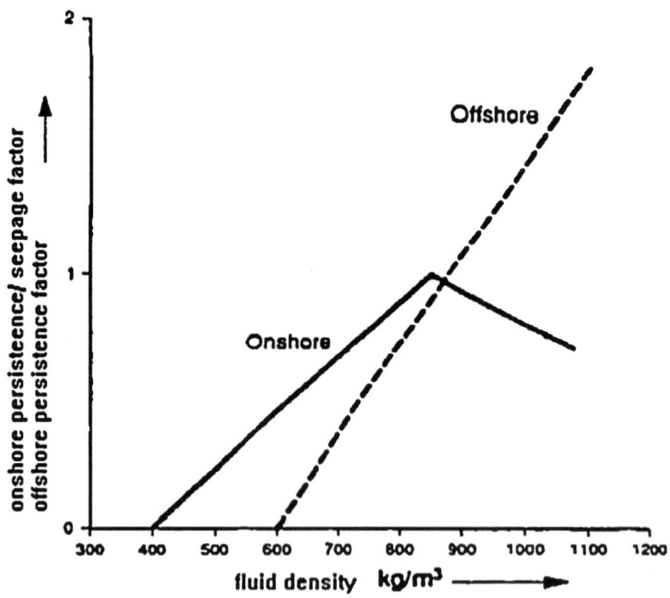

Fig. 3.7 Persistence factor *offshore* and persistence/seepage factor *onshore*

Table 3.10 Climate correction factor for persistence/seepage factor; E2

Climate	Average annual temperature (°C)	Correction factor, E2
Warm	>20	0.75
Moderate	5> and <20	1
Cold	>5	1.25

Table 3.11 Cleanup and/or other associated costs; E3 climate

Environment	Cleanup costs; E3 (US$/m^3)	Remarks
>40 km from shore	13	Based on surveillance of the released fluid only, while allowing self-degradation
5–40 km from shore	110 or 240	Value depends on remedial actions: Chemical dispersant treatment (dispersant and application) is 110 USD/m^3. Containment and recovery: (equipment, deployment, recovery, transport, and disposal) is 240 USD/m^3
<5 km from shore	3,500	Including coastal cleanup, fishing and tourism compensation, and amenity impact
Standard terrain	630	
Water course areas	2,200	
Designated environmentally sensitive areas	2,500	

Possible fines and other intangible costs related to the possible damage to the environment, e.g., loss of goodwill and credibility, are not included in the above costs. These should either be evaluated as a separate cost in an economic consequence evaluation or incorporated into the methodology by increasing the "cleanup and/or other associated costs" (E3) values. Actual costs are not included since they will be principal dependent.

The ECF of a leak is calculated from the formula

$$ECF = (L_m \times 1{,}000/R_o) \times L_e \times L_s \times E_1 \times E_2 \times \left(\frac{E_3}{100{,}000}\right). \qquad (3.6)$$

3.14 Leak Detection Techniques

Leak detection techniques are based on either continuous or intermittent measurements of specific parameters. Intermittent leak detection methods are often able to detect smaller leak rates compared with continuous leak detection techniques.

Some continuous techniques can only detect transient pipeline conditions during the onset of a leak and will not be able to identify the presence of a leak at a later time.

For some intermittent techniques, fluid transportation through the pipeline needs to be interrupted. Using intermittent techniques, the detection time of a leak will be completely dependent on the frequency of inspection.

Techniques for detection of leaks in liquid lines offer better performance than those for gas pipelines, which in turn are better than those for two-phase pipelines.

The conflicting balance of sensitivity to leaks and false alarms will determine the sensitivity setting of the leak detection system. Large leaks can normally be detected more rapidly than small ones. To maintain the user's confidence in the system, avoiding false alarms should have a higher priority than attempting to shorten the leak detection time or reducing the minimum detectable leak rate.

The performance of pipeline leak detection techniques is dependent on fluid type, operating pressure including fluctuations, batch or continuous operation, pipeline length and size, metering accuracy, etc.

To decide which technique to adopt depends on a detailed case-by-case evaluation. If the consequences of a leak are considered significant, then the more sophisticated techniques of leak detection are required. It may be necessary to deploy more than one leak detection technique in order to achieve the overall leak detection performance that is required.

Leak detection systems are categorized into the following groups according to their inherent principle of leak detection:

(a) Balancing of pipeline mass input versus output,
(b) Pressure and/or flow analysis,
(c) Monitoring of characteristic signals generated by a leak, and
(d) Off-line leak detection.

A summary of the capabilities and application of the various leak detection techniques is given in Table 3.12.

Leak rate categories used in Table 3.12:

Full bore rupture	100 % of flow ≥
Major leak	50–100 % of flow
Large leak	25–50 % of flow
Medium leak	5–25 % of flow
Small leak	1–5 % of flow

Table 3.12 Summary of the capabilities and application of leak detection techniques

Leak detection method	Leak type	Mode of operation	Response time	Leak location capability	Remarks
Low pressure	Gas: full bore ruptures	Any	Seconds to minutes	Offshore: none	Commonly used, high thresholds to avoid false alarms
	Liquid: major leaks			Onshore: Between block valves if pressure readings are available	
Pressure decrease/ flow increase	Gas: major leak	Steady state	Seconds to minutes	Offshore: none	–
	Liquid: large leaks			Onshore: Between block valves if pressure readings are available	
Pressure gradient along the pipeline	Gas: major leaks	Steady state	Minutes	Between block valves if pressure readings are available	Onshore only
	Liquid: Large leaks				
Negative pressure wave	Gas: large leaks	Steady state	Seconds to minutes	Within 1 km	Detects only the onset of a leak
	Liquid: medium leaks				
Wave alert	Gas: medium to large leaks	Steady and transient state	Seconds to minutes	Within 1 km depending on transducer spacing	Detects only the onset of a leak
	Liquid: small to medium leaks				
Mass balance	Medium to large leaks	Steady state	Minutes to hours	None	–
Corrected mass balance	Small to medium leaks	Steady and transient state	Minutes to hours	Offshore: none	–
				Onshore: between block valves	

(continued)

Table 3.12 (continued)

Leak detection method	Leak type	Mode of operation	Response time	Leak location capability	Remarks
Dynamic simulation	Small leaks	Steady and transient state	Minutes to hours	At best within 10 % of pipeline length	–
Statistical leak detection	Small leaks	Steady and transient state	Minutes to hours	Indication only	Low probability of false alarm
Ultrasonic leak detection pig	Liquids: small leaks (typical 50 l/h)	Intermittent	Depends on pigging frequency	Within 100 m	Hard liquids only
Acoustic reflectometry	Liquids: Large leaks (on-line), small to medium leaks (shutdown)	Steady state	Depends on monitoring frequency	Within 1 km	Hard liquids only
Differential static pressure test	Small leaks (hard liquids), medium leaks (soft liquids), large leaks for gas	During shutdown	Hours to days	Non, between block valves	Capabilities depend on length and temperature effects
Sniffer tube, hydrocarbon-sensing cables	All fluids, including multiphase: Small leaks	Any	Hours	Within 100 m	Short lines only

3.14.1 Balancing of Mass Input Versus Output

This category of leak detection systems relies on the fact that in a leak-free pipeline, the fluid mass flow into the pipeline equals the flow out. Using this mass balance principle, the flow-in and flow-out measurements are continuously monitored for any variations over a time interval. Volume flow readings should either be corrected for density or pressure and temperature variations to reference mass flows. To eliminate the effect of flow variations during normal operation, the flow readings should be averaged (totalized) over discrete time periods.

The uncorrected mass balance method can be applied only under steady-state operations as it does not allow for changes in the pipeline inventory, i.e., line pack variation. Its accuracy depends largely on the accuracy of the flow meters and on the steadiness of operations.

In addition to the inlet and outlet flow measurements, the corrected mass balance method uses a correction factor for any changes in the pipeline inventory. Pressure and, if necessary, temperature measurements at intervals along the pipeline are used for calculating the correction factor. The capability for detecting small leaks depends upon the number and accuracy of measurements along the length of the pipeline.

An alternative method is dynamic simulation, which is a model-assisted balance method. A real-time computer model calculates the inventory of the pipeline and the line pack variations of the pipeline under steady-state and transient operating conditions. It will correct not only for pressure and temperature effects but also for changes in fluid properties, such as where different batches of fluids are present in the pipeline at the same time. A difference between the mass balance predicted by the model and that actually measured indicates the presence of a leak. Also, unexpected flow and/or pressure trends are used as indicators of the occurrence of a leak.

The dynamic simulation method is similar to the corrected mass balance system. The main difference is that the dynamic simulation method calculates the pipeline inventory, whereas the corrected mass balance method interpolates between the measurements along the pipeline. The latter is usually considered to be less accurate because of the inherent accumulation of measurement errors.

The sensitivity of these methods is generally good. Their disadvantage is that they have limited capabilities for locating the leak.

To develop a statistical pipeline leak detection (SPLD), the system does not need complicated modeling of the pipeline inventory; it continuously calculates the statistical probabilities of a leak based on fluid flow and pressure measured at the inlet and outlet of a pipeline. Depending on the control and operation of a pipeline, the statistical technique is used to identify changes in the relationship between the pipeline pressure and flow which always occur when there is a leak.

The SPLD system works as a statistical filter, which is applied to a pipeline input/output balance and which decides between a leak-free and a leak-present hypothesis. Major advantages of this system are its simplicity and robustness compared with other software-based techniques. The SPLD system can run on a

PC and is capable of discriminating between fluctuations due to operational variations of the pipeline and the actual occurrence of a leak; it is thus very reliable for leak detection. The SPLD system has been commercialized since October 1991. The statistical filter of the SPLD system can also be combined with a commercially available dynamic simulation method, which makes the latter even more reliable. This combined statistical and dynamic simulation leak detection system is at present the most sophisticated leak detection system available.

3.14.2 Pressure and/or Flow Analysis

The operation of a pipeline can be characterized by the flow of the fluid and the pressure gradient along the pipeline. Pressure drop and flow along a pipeline are related to the flow resistance of the pipeline. A leak will alter the pressure drop profile of a pipeline and therefore affect the "normal" pressure and flow relationships. Detection of such alterations can be used to indicate the occurrence of a leak.

If a large leak occurs, particularly in the upstream part of a pipeline, the inlet pressure will drop. Observation of a lower-than-expected inlet pressure indicates the presence of a leak. Detection of low pressure is usually connected to an automatic shutdown system. To avoid false alarms, the system is usually set such that only major leaks can be detected.

A leak will result in an increase in flow upstream and a decrease in flow downstream of the leak. As a result of this, the pressure gradient will increase upstream and decrease downstream of the leak. The occurrence of a discontinuity in the pressure gradient, which is calculated from the pressure readings along the pipeline, is an indication of a large leak. The rate of change of pressure and flow readings can also be monitored and used to detect sudden changes which indicate the occurrence of a leak.

The combined pressure decrease/flow increase method uses the fact that a leak in an operational pipeline will cause an increase in the flow and a decrease in the pressure upstream of the leak. The simultaneous occurrence of both is an indication of a leak.

3.14.3 Monitoring of Characteristic Signals Generated by a Leak

A suddenly occurring leak will cause a sudden pressure drop at the leak location in the pipeline. This sudden pressure drop will create a pressure wave traveling at sonic velocity both upstream and downstream from the leak. Detection of this pressure wave is an indication of the occurrence of a leak. The response time of this negative pressure wave technique is very short because it responds to waves that travel at sonic velocities (in crude oil, approximately 1,000 m/s). When the wave is detected both upstream and downstream of the leak, the location of the leak may be calculated from the time difference of detection by the nearest sensors on either side of the leak location. The system will only respond to an instantaneously occurring leak of measurable size. In practice, the sensitivity can be poor because the alarm thresholds are often set high to avoid false alarms triggered by pressure transients generated by upstream or downstream processing plant or other noise producing installation, such as pump or compressor stations.

A system which is less sensitive to pipeline noise than the negative pressure wave system uses dual transducers which filter out noise signals. The system is made directional, i.e., it detects signals originating from either the upstream or the downstream direction of the pipeline. This is achieved by installing the two transducers at an appropriate from each other and using an electronic signal subtracting system.

Leak detection based on negative pressure wave techniques will only detect the initiation of a leak and not its presence. If the pressure wave created at the moment of leak initiation is not detected, the leak will not be noticed.

Liquid escaping under pressure through a small opening produces supersonic noise. An ultrasonic leak detection pig, which is equipped with hydrophones and data recording, can detect and locate the presence of a leak. A very small leak, down to 10 l/hr, can be detected and fairly accurately located with this technique. Being intermittently operated, the response time will depend on the frequency of running the ultrasonic leak detection pig.

A hydrocarbon-permeable tube (sniffer tubes) can be laid in close proximity along the pipeline. Small leaks of hydrocarbons from the pipeline which have permeated into the tube will be detected when the tube is periodically purged into a gas analyzer.

Hydrocarbon-sensing cables can be laid along the pipeline. Electrical properties of the cable change when hydrocarbons come in contact with the cable. Contact with water does not affect the properties of the cable.

A prototype system for the measurement of methane in sea water has been developed. The device, which is mounted on a ROV, extracts dissolved gas from a continuous flow of water and determines the methane content using infrared absorption techniques.

Remote sensing of hydrocarbon emissions, e.g., using an infrared technique from an aircraft, is becoming commercially available. Particularly for gas and multiphase pipelines, this offers a powerful alternative to ground-based patrolling techniques.

3.14.4 Off-Line Leak Detection

Intelligent pigs have been developed for detection and location of leaks in a pipeline using flow direction recognition in a blocked-in pressurized pipeline. This bidirectional pig has an opening through the body with a sensitive flow meter and a transmitter. By locating the pig at various points along the line and using aboveground interpretation of the flow measurements through the pig, the leak can eventually be located. Locating the leak, however, is time-consuming, and the line should be equipped with pumping or pressurizing facilities at both ends. This system is of interest for pipelines larger than 8 inches in diameter when a small leak has been detected, but its location is unknown.

An alternative to the above technique for pipelines smaller than 8 inches is a bidirectional pig equipped with a differential pressure transducer and a transmitter. When located in the pipeline, the pig measures the pressure drop on either side. The leak will be on the side at which the pressure drops more rapidly.

The pressure in a blocked-in pressurized pipeline will drop when there is a leak. For a static pressure leak test, the pipeline, or a section of it, is pressurized with the transported hydrocarbon fluid to the MAOP. If pressurizing to a higher level is required, the leak test shall be done with water for safety and environmental reasons. After pressurizing, the block valves are closed and the pressure and temperature are monitored for a specified period of time (24 h minimum). A differential static pressure test can be carried out if block valves are equipped with differential pressure transducers. A difference in the rate of pressure drop in two adjacent sections that cannot be explained by temperature effects, inaccuracy of readings or valve leakage is an indication of a leak. There are uncertainties about the advantages and disadvantages of pressure testing existing pipelines for condition monitoring purposes at pressure higher than the MAOP.

The advantage of pressure testing at high pressures for leak detection is that an existing leak is detected more easily. Also, long defects which have almost broken the surface can be opened, resulting in a leak which is also detected. The disadvantage is the risk that existing defects might be enlarged and/or activated to grow, possibly leading to failures during normal pipeline operations following the pressure test.

(Note: Pressure testing above the MAOP is primarily done for strength testing in order to avoid a pipeline rupture (DEP 31.40.40.38-Gen).

Sound which is generated when liquid is forced through a small opening during pressure testing can be detected by acoustic monitoring. For pipelines transporting hard liquids, leak detection by an acoustic reflectometry method is feasible. The technique is based on the phenomenon that a pressure wave traveling through a pipeline is reflected at the position of a leak, due to a local change of acoustic properties. For lines which are used intermittently, this technique can be used during downtime when the level of disturbing noise is low.

3.15 Leakproof Control of Pipelines, Gas Piping, Tanks, and Technological Installations Using Radioactive Tracers

The tracer can be added to the pressurizing fluid for detection of small leaks. The leak is detected by patrolling the pipeline with a detector which is sensitive to the tracer or by visual observation of a visible tracer.

The method of isotopic tracers can be used as a complementary technique to the hydraulic and pneumatic pressure tests. The method shows many advantages. It is sensitive and easy to apply. The test and preliminary works are not time-consuming. In addition, the isotopic test can be carried out at low pressure (0.2–0.4 MPa) without excessive effort of construction materials. It should be mentioned that hydraulic tests cannot be applied to some object despite the fact that such tests are necessary.

The method is based on the phenomenon of migration of the isotopic tracer in the direction of the leak, its penetration to the surrounding medium, adsorption of the tracer onto this medium, and measurement of gamma radiation emitted by the tracer. Methyl bromide labeled with radioisotope 82Br is the best tracer.

The chemical reaction is carried out in a special generator constructed for this purpose. Depending on the type of the generator, activities up to 10 Ci (370 G Bq) can be handled. For transportation of the gaseous radioactive tracer to the dispensing point, special containers are used. The radiotracer is sucked off from the generator to the container. The containers vary in size depending on the activity of the tracer to be transported.

The detection and localization of leaks are carried out by recording changes in radiation intensity (resulting from adsorption of the radioactive tracer in the insulating material) with the aid of portable radiometers with scintillation detectors.

The measurement of the dose rate on the surface of the generator container and the determination of the boundaries of the controlled area are carried out with the aid of Geiger–Muller counters. When simultaneous measurements are carried out at many points, highly sensitive multichannel radiometers are used.

3.15.1 Leak Detection in Pipelines for Liquids

In methyl bromide (and in the case of water–potassium bromide) labeled with radioactive 82Br should be introduced to the liquid in the pipeline and detected

Fig. 3.8 Pipeline pig

with the aid of a special gamma-ray detector, placed in a pressure proof pig (Fig. 3.8) moving with the medium through the pipeline. The detector is introduced to the pipeline after the tracer has passed. It continuously measures the natural radiation in the pipeline and also the peaks of gamma radiation of 82Br (if present). The minimum activity which must be present in the leak in order to be detected is 1–10 μ Ci (37–370 k Bq).

The signals recorded on the tape are called "general localization" of the leak and supply information as to the position of the leak with an accuracy of several to several tens of meters depending on the distance between the distance markers (^{60}Co sources placed on the outer walls of the pipeline) and on the rate of tape winding of the recorder. Exact localization of the leak is carried out by searching the radiation of the ground above the pipeline in the area delineated as described above. In the case of pipelines of very small diameters, to which radiometers cannot be introduced, only accurate localization is carried out.

In such cases, the detection is based on the measurement of tracer radiation that penetrated through the leak to upper layers of the ground by means of a detector moved just above the surface of the ground. In this method, it is necessary to stop the operation of the pipeline for some time in order to follow the movement of the added tracer. The total applied activity should be 20 Ci (740 G Bq). The minimum leak detected shall be in the range of 30–1,000 cm^3/h. With the detectors introduced into the pipeline, it is possible to examine pipelines with diameters of the range 200–800 mm.

3.15.2 Leak in Gas Pipelines Detection

The choice of the method of examination depends on the size of the leak, the length of the section of the gas pipeline to be examined, and the pipe diameter. These factors affect the rate of movement of the tracer on the way from the point of its introduction to upper layers of the soil in the leak zone, where the tracer is detected by means of a gamma-ray detector.

Leak testing with the aid of radioactive tracers can be carried out before, during, and after the pressure tests for strength. The choice of the variant of the method depends on the pressure decrease in the conventional pressure test.

3.15.3 The Method of Labeling Total Volume of the Gas Pipeline

The method can be applied to short pipelines and small leaks. The medium and the tracer should be pumped into the pipeline at one or more points in such a way as to evenly distribute the tracer in the whole volume of the line.

For the detection, apply high test pressure as much as possible. When a definite pressure is attained, the pumping is interrupted. After several hours (this time is necessary to allow the labeled gas to pass through the smallest leaks to the upper layers of the ground), the radiation of the tracer in the leak is recorded. In this method, it is possible to simultaneously detect all leaks without the necessity of following the movement of the tracer, i.e., without digging up the pipeline when locating leaks.

3.15.4 Methods of Single Injection of the Tracer

The method is used for testing short sections. The gas should be pumped up to a definite pressure.

The pumping is then interrupted, and the radioactive tracer should be introduced into the middle section of the gas piping.

Introduction of the tracer into the middle section of the gas piping permits to reduce the time of localization of the leaks. Radiometers located on pipe on both sides of the point of injection of the tracer permit to determine the direction of movement of the tracer, i.e., the part in which the leak occurs.

3.15.5 Method of Injection of the Tracer at Several Points of the Gas Pipeline

The method is used in the case of examination of long sections of gas pipeline. The leak is localized by following the movement of the radioisotope with the aid of radiometers located on both sides of the injection point. In this method, very small portions of the tracer are used (the radioactivity of each is about 1 m Ci). The last portion of the tracer injected in the close vicinity of the leak is sufficiently high to be measured above the leak zone.

3.15.6 Method of Tracer Injection Under Constant Pressure of the Inspection Medium

The method can be applied to gas pipelines of any length with large leaks. It should be noted that the single pumping may not be sufficient to transport the tracer to the leak zone; the gas is continuously pumped under a constant pressure. Continuous pumping of the gas shows advantages in the case of small leaks since it permits to reduce the time of their localization. It is necessary to continuously maintain definite supply of the gas in order to compensate for the losses through the leaks. The localization can be carried out at any place, and the minimum detectable leakage rate is less than 30 cm^3/h.

3.16 Emissions to Land

To characterize emissions to land, three general EETs are discussed here:

- groundwater monitoring
- spills

3.16.1 Groundwater Monitoring

Some facilities conduct monitoring of groundwater to characterize releases from the facility. Where available, this monitoring data may be used to assist in the characterization of releases. This involves determining upstream and downstream concentrations and using this information in conjunction with groundwater flow information to determine the contribution of the facility to pollutant levels in the groundwater.

This approach is reasonable in situations where there is no loss of substances (e.g., due to evaporation) prior to the substance entering the groundwater and where the time between the release occurring and the substance entering the groundwater is minimal. Therefore, for those facilities where groundwater monitoring captures all releases to land, such monitoring can be used as a reasonable measure of emissions to the environment. If this is not the case (e.g., where the rate of transmission through the soil/clay is low or where there are other routes whereby substances to land are carried off-site, such as evaporation or surface runoff), it will be necessary to characterize such releases using the other EETs presented in this section.

3.16.2 Spills

For many facilities, the primary source of releases will be through spills (this may also include intentional spillage due to vessel washdown). Accidental spills can contribute to releases to land (directly), to water (through runoff), and to air.

As discussed above, unless the spilled material is routed to a secure containment facility, the quantity of material spilled, less than the quantity collected (or cleaned up), is required to be reported under the NPI. In practical terms, a log of spillages could be maintained detailing the quantities spilled and the composition of the spill (in particular, the quantities of NPI substances spilled).

This log could then form the basic information required to meet NPI reporting requirements.

The quantity spilled can be partitioned into air emissions and releases to land by assuming that all of the light end fraction is volatilized and the remaining fraction is released into the ground. The time, quantity of spill, temperature, and porosity of the soil all play an important part in the estimation of release. The evaporation rate of compounds into the atmosphere is given by the following equation

$$E_i = 1.2 \times 10^{-10} \times (M \times poi/T) \times u^{0.78} \times y^{0.89} \tag{3.7}$$

where:

Ei Evaporation rate of substance "i" (g/s);
u Wind speed over the surface of the spill (cm/s);
x Downwind dimension (cm);
y Crosswind dimension (cm);

M Molecular weight of the substance;
poi Vapour pressure of substance "i" at spill temperature T (dyne/ $cm^2 = 0.0001$ kPa); and
T Temperature (K).

Once losses to the atmosphere have been quantified, releases to land can be estimated using the following equation

$$\text{ERLAND, i} = \text{QttySPILL} - ((\text{time}) \times (\text{Ei})) \tag{3.8}$$

where:

ERLAND, I The emission to the land of compound "i";
QttySPILL The quantity of compound in the liquid spilled;
Ei The loss through evaporation of substance "i" as estimated using the evaporation equation above; and
Time The time period between which the liquid was initially spilled and eventual clean up.

3.17 Evaluation of Soil Infiltration Problems

Sodium (Na) is one of the most problematic ions present in excessive concentrations and causes specific toxicity. In addition, another indirect effect of high sodium content is the deterioration of the physical condition of soil such as formation of crusts, water logging, and reduced soil permeability. If the infiltration rate is greatly reduced, it may be impossible to supply the crop or landscape plant with sufficient water for good growth. In this section a simple predictive tool, which is easier than existing approaches and less complicated with fewer computations, is formulated to accurately predict the sodium adsorption ratio (SAR) as a function of concentration of Na^+, magnesium (Mg^{2+}), calcium (Ca^{2+}), salinity of applied water, and the ratio of bicarbonate (HCO_3^-) over Ca^{2+} for interpretations of water quality for irrigation.

The proposed method showed consistently accurate results for salinity of applied water up to 8 dS/m and ratio of HCO_3^- over Ca^{2+} up to 20. Predictions showed an excellent agreement with the reported data with an average absolute deviation less than 3 %.

3.17.1 Developed Predictive Tool

The water infiltration problem occurs within the top few centimeters of the soil and is mainly related to the structural stability of the surface soil. To predict a potential infiltration problem, the SAR is often used.

$$SAR = \frac{Na^+}{\sqrt{\frac{Ca^{2+} + Mg^{2+}}{2}}} \tag{3.9}$$

where the cation concentrations are expressed in meq/L. The adjusted SAR (R_{Na}) is a modification of Eq. 3.1 which takes into account changes in calcium solubility in the soil water. Use of R_{Na} value is preferred in irrigation applications with reclaimed water because it reflects the changes in calcium in the soil water more accurately. At a given SAR, the infiltration rate increases as salinity increases or decreases as salinity decreases.

$$adj\, R_{Na} = \frac{Na^+}{\sqrt{\frac{Ca_x^{2+} + Mg^{2+}}{2}}} \tag{3.10}$$

where Na^+ and Mg^{2+} concentrations are expressed in meq/L, and the values of Ca_x^{2+} also expressed in meq/L is obtained through novel predictive tool presented in this work. Therefore, R_{Na} and electrical conductivity (EC_w) of irrigation water should be used in combination to evaluate the potential permeability.

Reclaimed water is normally high in calcium, and there is little concern for the water dissolving and leaching too much calcium from the surface soil. However, reclaimed water is sometimes high in sodium; the resulting high SAR is a major concern in planning irrigation project with reclaimed water.

The required data to develop this method include the reliable data [57] to predict Ca_x^{2+} parameter in Eq. 3.2 to reflect the changes in calcium in the soil water more accurately, as a function of salinity of applied water (E_c) and ratio of HCO_3^-/Ca^{2+}, R, in meq/L. In this work, the Ca_x^{2+} parameter is predicted rapidly by proposing a simple tool.

So, Eq. 3.8 represents the proposed governing equation in which four coefficients are used to correlate the Ca_x^{2+} parameter values as a function of ratio of HCO_3^-/Ca^{2+} (R) for different salinities of applied water (E_c) where the relevant coefficients have been reported in Table 3.13.

$$\ln\left(Ca_x^{2+}\right) = a + \frac{b}{R} + \frac{c}{R^2} + \frac{d}{R^3} \tag{3.11}$$

where:

$$a = A_1 + B_1 E_c + C_1 E_c^2 + D_1 E_c^3 \tag{3.12}$$

$$b = A_2 + B_2 E_c + C_2 E_c^2 + D_2 E_c^3 \tag{3.13}$$

$$c = A_3 + B_3 E_c + C_3 E_c^2 + D_3 E_c^3 \tag{3.14}$$

Table 3.13 Tuned coefficients used in Eqs. 3.4–3.7 for estimating absorption efficiency according to reliable data [57]

Coefficient	Tuned coefficients for $\frac{HCO_3^-}{Ca^{2+}}$ ratio less than 1	Tuned coefficients for $\frac{HCO_3^-}{Ca^{2+}}$ more than 1
A_1	$2.611975700671 \times 10^{-1}$	-1.2613155584148
B_1	$1.5323734035744 \times 10^{-1}$	$-2.655970585552 \times 10^{-1}$
C_1	$-2.5301512329647 \times 10^{-2}$	$5.8283608290452 \times 10^{-2}$
D_1	$1.5711688598023 \times 10^{-3}$	$-3.5562691856642 \times 10^{-3}$
A_2	$4.4312371304785 \times 10^{-1}$	7.237129012651
B_2	$-1.33878731261208 \times 10^{-4}$	$1.0837184119463 \times 10^{-1}$
C_2	$1.33146329099254 \times 10^{-4}$	$-3.4537210697001 \times 10^{-2}$
D_2	$-1.4920964960815 \times 10^{-5}$	$2.4184818439973 \times 10^{-3}$
A_3	$-3.57929622902405 \times 10^{-2}$	-9.6247853800639
B_3	$1.5429385602988 \times 10^{-5}$	$-2.3481099697109 \times 10^{-1}$
C_3	$-1.4841265770122 \times 10^{-5}$	$7.4343560147113 \times 10^{-2}$
D_3	$1.6517869818181 \times 10^{-6}$	$-5.1878460983131 \times 10^{-3}$
A_4	$9.7273165732822 \times 10^{-4}$	4.636747565375
B_4	$-5.5936769324377 \times 10^{-7}$	$1.4185005340368 \times 10^{-1}$
C_4	$4.7062585055241 \times 10^{-7}$	$-4.4690499765614 \times 10^{-2}$
D_4	$-5.1248968401358 \times 10^{-8}$	$3.1104959601422 \times 10^{-3}$

$$\mathrm{d} = A_4 + B_4E_c + C_4E_c^2 + D_4E_c^3 \tag{3.15}$$

Below is a summary of the list of symbols for equations of presented Sect. 3.17:

A	Tuned coefficient
B	Tuned coefficient
C	Tuned coefficient
D	Tuned coefficient
Ca^{2+}	Ca concentration, meq/L
Na^+	Na concentration, meq/L
Mg^{2+}	Mg concentration, meq/L
E_c	Salinity of applied water, dS/m
R	Ratio of HCO_3^-/Ca^{2+}
R_{Na}	Adjusted SAR
SAR	Sodium adsorption ratio

These optimum tuned coefficients (A, B, C and D) help to cover the Ca_x^{2+} parameter value data reported in the literature [57].

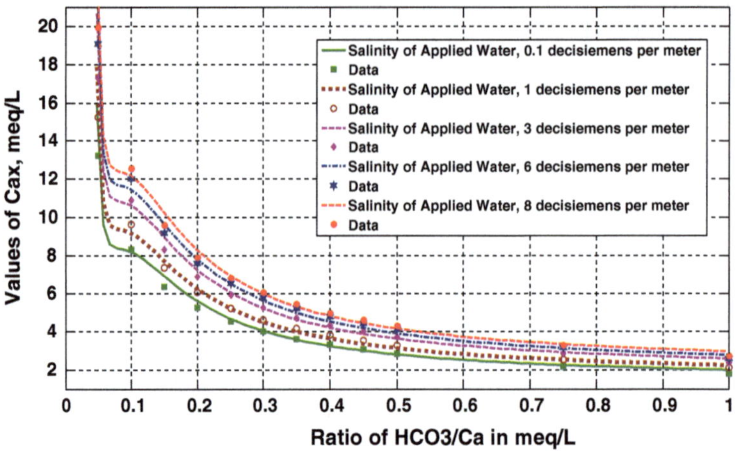

Fig. 3.9 Prediction of Ca_x^{2+} in comparison with reported data [57] for ratio of HCO_3^-/Ca^{2+}, less than 1. (Reprinted with permission [92])

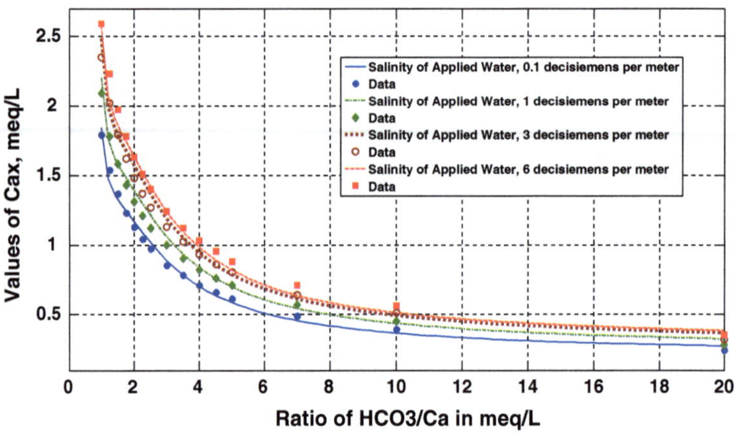

Fig. 3.10 Prediction of Ca_x^{2+} in comparison with reported data [57] for ratio of HCO_3^-/Ca^{2+}, greater than 1 (Reprinted with permission [92])

This predictive tool is recommended for salinity of applied water (E_c) up to 8 dS/m and ratio of HCO_3^-/Ca^{2+} up to 20. In order to determine whether an infiltration problem will develop or not:

Guidelines for interpretation of water quality for irrigation table are reported in [57]. The coefficients presented in Table 3.13 can be retuned quickly according to

Table 3.14 Given data in example

Water quality parameter	Concentration, mg/L
BOD	39
TSS	160
Total N	4.4
Total P	5.5
pH	7.7
Ca^{2+}	37
Mg^{2+}	46
Na^{+}	410
K^{+}	27
HCO_3^-	295
SO_4^{2-}	66
Cl^-	526
Boron	1.2
Electrical conductivity (dS/m)	2.4
TDS	1,536
Alkalinity	242
Hardness	281

the proposed approach if new data become available in the future just by readjusting the tuned coefficients.

Figures 3.9 and 3.10 show the results of the proposed predictive tool in comparison with the reported data [57], and Pettygrove, G.S., Asano 1985) to predict Ca_x^{2+} parameter in Eq. 3.2 to reflect the changes in calcium in the soil water more accurately, as a function of salinity of applied water (E_c) and ratio of HCO_3^-/Ca^{2+}, R, in meq/L. It is clear that the proposed method yields results with good accuracy.

Typical example is given below to illustrate the simplicity associated with the use of proposed correlation for rapid estimating Ca_x^{2+} parameter to calculate adsorbed sodium ratio.

3.17.2 Sample Calculations for Practice Engineers

The following water quality analyses in Table 3.14 were reported on an aerated lagoon effluent which will be used for irrigating agriculture land. Using the reported water quality data, (1) calculate R_{Na} and (2) determine whether an infiltration problem may develop by using this effluent for irrigation.

- **Solution**

Firstly, convert the concentrations of the related water quality parameters to meq/L:

$$\mathrm{Ca}^{2+} = 37/20.04 = 1.85\ \mathrm{meq/L}$$

$$\mathrm{Mg}^{2+} = 46/12.15 = 3.79\ \mathrm{meq/L}$$

$$\mathrm{Na}^{+} = 410/23 = 17.83\ \mathrm{meq/L}$$

$$\mathrm{HCO}_3^{-} = 295/61 = 4.84\ \mathrm{meq/L}$$

Determine the value of Ca_x^{2+} using the given water quality data.

Salinity of applied water, E_c=2.4 dS/m

Ratio of $\mathrm{HCO_3}^-/\mathrm{Ca}^{2+} = 4.84/1.85 = 2.62$

Because Ratio of $\mathrm{HCO_3}^-/\mathrm{Ca}^{2+}$ or R is more than 1, so we use the second column coefficients:

a	−1.36211957 (from Eq. 3.9)
b	7.276462851 (from Eq. 3.10)
c	−9.710091705 (from Eq. 3.11)
d	4.6883179939 (from Eq. 3.12)
Ca_x^{2+}	1.29859 (from Eq. 3.8)

Calculate the R_{Na} using Eq. (3.2)

$$adj\,R_{Na} = \frac{Na^+}{\sqrt{\frac{Ca_x^{2+}+Mg^{2+}}{2}}} = \frac{17.83}{\sqrt{\frac{1.29+3.79}{2}}} = 11.19 \tag{3.16}$$

The calculated R_{Na} (11.19) has good agreement with reported R_{Na} (11.29) where the deviation is less than 1 %.

In order to determine whether an infiltration problem will develop or not:

According to guidelines for interpretations of water quality for irrigation with R_{Na} (11.19) and salinity of applied water (2.4 dS/m), no restrictions are indicated for the use of this reclaimed water.

Noise Pollution Control

4

Abstract

In simple terms, noise is unwanted sound. Noise can disturb human's work, rest, sleep and communication, and it can damage hearing. Indeed, the most important issue is industrial noise problem and a need for noise control and hearing conservation programs. This chapter provides guidelines comprising: (a) Brief explanation of the principle and essential standards for noise emitted by machineries and equipment. (b) Brief explanation of the principle underlying standards for noise in workplace areas, and noise control procedures in plants/ complexes, offices, conference rooms, etc. (c) Brief explanation of the principle for vibration control.

Keywords

Acoustical barrier · Acoustical design · Damping · Decibel · Noise · Noise limits · Pollution · Silencing equipment · Sound · Vibration control

Sound is a form of energy which is emitted by a vibrating body and on reaching the ear causes the sensation of hearing through nerves. Sounds produced by all vibrating bodies are not audible. The frequency limits of audibility are from 20 to 20,000 Hz.

Sounds of frequencies less than 20 Hz are called infrasonic and greater than 20,000 Hz are called ultrasonics. Since noise is also a sound, the terms noise and sound are synonymously used and are followed in this chapter.

A noise problem generally consists of three inter-related elements—the source, the receiver, and the transmission path. This transmission path is usually the atmosphere through which the sound is propagated, but can include the structural materials of any building containing the receiver (see Fig. 4.1).

This chapter is prepared in two parts:

- Part 1 Noise Control
- Part 2 Vibration Control

A. Bahadori, *Pollution Control in Oil, Gas and Chemical Plants*,

DOI: 10.1007/978-3-319-01234-6_4,

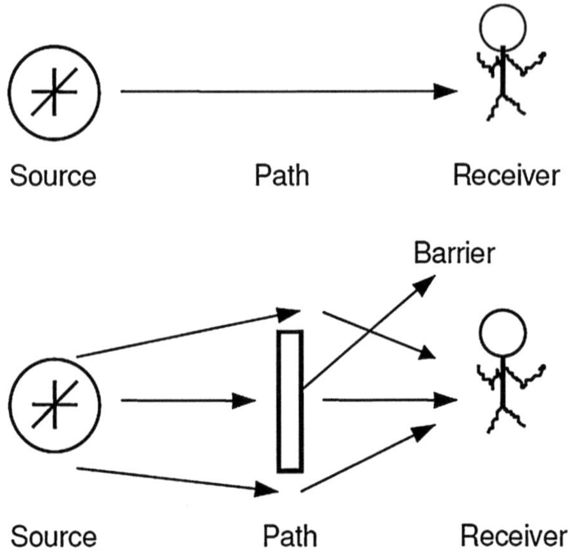

Fig. 4.1 Inter-relationship between the elements of noise

This chapter defines the procedures for noise control of plant and equipment. It specifies how to derive maximum allowable noise levels for equipment installed in oil refineries, chemical plants, gas plants and, where applicable, in exploration and production facilities and supply/marketing installations.

It applies equally to design and construction of new plants and to modification of existing plants.

The actual noise limits for the equipment under consideration shall be specified in the equipment requisitions under guidance of the management and appropriate medical and safety authorities.

Part 1 of this chapter also deals with the particular requirements for installing the acoustic insulation and is intended for the noise control engineer to consider the acoustic design of the plant.

Part 2 of this chapter defines the vibration control of particular equipment.

4.1 Basic Theory and Calculations

The intensity of sound is measured in sound pressure levels (SPL) and common unit of measurement is decibel, (dB). The community (ambient) noise levels are measured in the A-weighted SPL, abbreviated dB(A). This scale resembles the audible response of human ear. Sounds of frequencies from 800 to 3,000 Hz are covered by the A-weighted scale. If the sound pressure level, L_1, in dB is measured at r_1 meters, then the sound pressure level, L_2, in dB at r_2 meters is given by,

$$L_2 = L_1 - 20 \log_{10}(r_2/r_1) \tag{4.1}$$

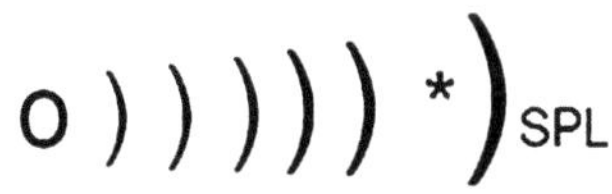

Fig. 4.2 Definition of sound pressure

Example:

If the distance from a noise source is doubled, find out the noise levels.

Solution:

Given, $r_2 = 2r_1$

We have,

$$L_2 = L_1 - 20\log_{10}(r_2/r_1)$$

Substituting, we get,

$$\begin{aligned} L_2 &= L_1 - 20\log_{10}(2r_1/r_1) \\ &= L_1 - 20\log_{10}(2) \end{aligned}$$

$$\begin{aligned} \text{i.e., } L_2 &= L_1 - 20 \times 0.301 \\ &= L_1 - 6.02 \end{aligned}$$

i.e., the noise level will decrease by 6 dB for doubling of distance from the source.

If the sound levels are measured in terms of pressure, then, sound pressure level, LP is given by,

$$L_P = 20\log_{10}(P/P_0)\ \text{dB(A)} \tag{4.2}$$

The LP is measured against a standard reference pressure, $P_0 = 2 \times 10^{-5}\ \text{N/m}^2$ which isequivalent to zero dBs. The sound pressure is the pressure exerted at a point due to a sound producing source (see Fig. 4.2).

4.1.1 Addition of Sound Levels

The effective sound levels form two or more sources cannot be simply added algebraically. For example, the effective sound level from two air conditioners

Table 4.1 Addition of sound levels, L_1 and L_2 ($L_1 > L_2$)

L_1–L_2, dB	Add to L_1 (dB)
0 or 1	3
2 or 3	2
4–8	1
9 or more	0

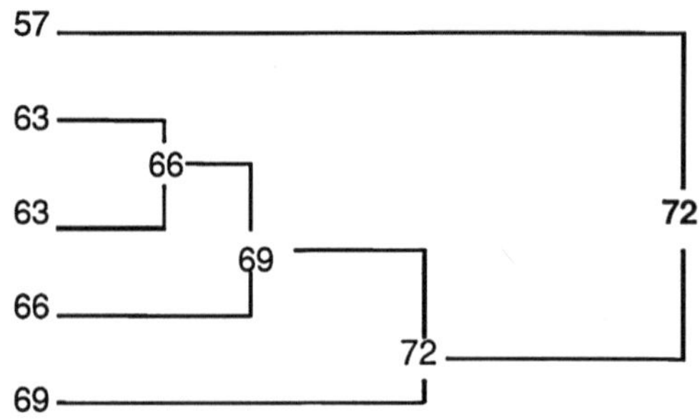

Fig. 4.3 A schematic for effective sound-level computations

60 dB(A) each, say is not 60 + 60 = 120 dB (A) but 60 + 3 = 63 dB(A) (see Table 4.1). Similarly, the effective sound level of 57, 63, 63, 66, and 69 dB is 72 dB. The computation is illustrated below in Fig. 4.3.

4.2 Frequency Analysis

The frequency analysis allows to separate the main components of the signals by dividing the frequency range of interest into smaller frequency bands using a set of filters. We may distinguish between noises that consist of regularly repeated or periodic sounds and those that consist of aperiodic sounds.

The simplest periodic sound is a pure tone, i.e., a pressure disturbance that fluctuates sinusoidally at a particular frequency. The lower the frequency, the longer is the wave length (wavelength = velocity of sound/frequency).

The noise produced by most sources of community noise, such as automobiles or aircraft engines, are examples of aperiodic sounds. Such sounds cannot be subdivided into sets of harmonically related pure tones but can be described in terms of components extending over finite frequency bands. Such frequency analysis are often done in bands of octaves or 1/3 octaves.

An octave band is a frequency band with upper and lower cutoff frequencies having a ratio of 2. The cut off frequencies of 707 and 1,414 Hz define an octave band, whose band center frequency is 1,000 Hz and would be referred to as the 1,000 Hz octave band. Frequency analyzers can be divided into two groups, viz. constant band width analyzer and constant percentage bandwidth analyzer.

In the constant bandwidth analyzer, the filter bandwidth is kept constant throughout the frequency range while in the constant percentage bandwidth analyzer, the bandwidth is proportional to the center frequencies.

The constant percentage bandwidth analyzer is widely used. The nine preferred center frequencies for noise-level measurement are 31.5, 63, 125, 250, 500, 1,000, 2,000, 4,000, and 8,000 Hz.

As already mentioned, dB(A) values give emphasis to sounds in the range of about 800–3,000 Hz. Since the sound-generating frequencies are not fully covered under dB(A), for detailed evaluation and engineering design, the multiple-number descriptions provided by frequency analysis are often required.

4.3 Noise Measurement Instruments

Noise measurement is an important diagnostic tool in noise control technology. The objective of noise measurement is to make accurate measurement which gives us a purposeful act of comparing noises under different conditions for assessment of adverse impacts of noise and adopting suitable control techniques for noise reduction. The various equipment used for noise-level measurement are summarized at Table 4.2.

The principle and the components of noise-measuring instruments is summarized below.

A sound-level meter consists basically of a microphone and an electronic circuit including an attenuator, amplifier, weighting networks or filters, and a display unit. The microphone converts the sound signal to an equivalent electrical signal. The signal is passed through a weighting network which provides a conversion and gives the sound pressure level in dB.

The instructions laid down by the noise-level meter manufacturers shall be followed while using the instruments.

The time constants used for the sound-level meter standards are:
S (Slow) = 1 s
F (Fast) = 125 ms

Relatively steady sounds are easily measured using the "fast" response and unsteady sounds using "slow" response. When measuring long-term noise exposure, the noise level is not always steady and may vary considerably, in an irregular way over the measurement period. This uncertainty can be solved by measuring the continuous equivalent level, which is defined as, the constant sound pressure level which would have produced the same total energy as the actual level

Table 4.2 Equipment used in the measurement of noise levels

Equipment	Specification/area of usage
Sound-level meter	Type 0: Laboratory reference standard
	Type-1: Lab use and field use in specified controlled environment
	Type 2: General field use (commonly used)
	Type 3: Noise survey
Impulse meters	For measurement of impulse noise levels, e.g., hammer blows, and punch press strokes etc
Frequency analyzers	For detailed design and engineering purpose using a set of filters
Graphic recorders	Attached to sound-level meter. Plots the SPL as a function of time on a moving paper chart
Noise dosimeters	Used to find out the noise levels in a working environment. Attached to the worker
Calibrators	For checking the accuracy of sound-level meters

over the given time. It is denoted as **Leq**. The display of Leq facility is also available in certain models of sound-level meters. This is the desired parameter for assessment of ambient noise levels.

4.4 Noise Control

The control of noise in a plant is required for the following reasons:

- To conserve the hearing of personnel;
- To reduce speech and work interference;
- To provide quiet accommodation for personnel;
- To prevent annoyance to the neighboring community.

Noise limits can be given for each of the above aspects of noise control, relative to certain areas inside or outside the plant.

Noise limits for each item of equipment shall be derived from the general noise limits, as applicable for the equipment in its specific position in the plant, which are commonly referred to as "equipment noise limits."

Each potential noise source shall be subject to the requirements of this specification.

The specified limits shall be met for the design operating conditions of the plant and for other operating conditions which occur occasionally, such as start-up, shutdown, regeneration, and maintenance. During construction, noise levels shall not exceed the limits applicable for operation of the completed plant.

Community noise problems are solved through the elimination or prevention of annoyance to neighbors and/or reduction in noise levels to within the standard limits.

4.4.1 Sound and Decibels

Sound is a disturbance which propagates through a medium having the properties of inertia (mass) and elasticity. The medium by which audible sound is transmitted is air.

The dB is a logarithmic unit that indicates the ratio of a physical quantity (usually power or intensity) relative to a specified or implied reference level. A ratio in dBs is ten times the logarithm to base 10 of the ratio of two power quantities [1]. A dB is one tenth of a bel, a seldom-used unit named in honor of Alexander Graham Bell.

The dB is used for a wide variety of measurements in science and engineering, most prominently in acoustics, electronics, and control theory. In electronics, the gains of amplifiers, attenuation of signals, and signal-to-noise ratios are often expressed in dBs. The dB confers a number of advantages, such as the ability to conveniently represent very large or small numbers, and the ability to carry out multiplication of ratios by simple addition and subtraction.

4.4.2 Intensity and Power

As with any energy source, the power of an acoustic source may be expressed in watts. In a free field, sound will propagate from a non-directional source in a spherical manner. Since the area of a sphere increases proportionately with the increase in distance from the source, the intensity, or power per unit area, decreases because the power is being distributed over an expanding area.

Often neglected, noise induces a severe impact on humans and on living organisms. Some of the adverse effects are summarized below.

Annoyance: It creates annoyance to the receptors due to sound-level fluctuations. The aperiodic sound due to its irregular occurrences causes displeasure to hearing and causes annoyance.

Physiological effects: The physiological features like breathing amplitude, blood pressure, heart-beat rate, pulse rate, and blood cholesterol are effected.

Loss of hearing: Long exposure to high sound levels causes loss of hearing. This is mostly unnoticed, but has an adverse impact on hearing function.

Human performance: The working performance of workers/human will be affected as they will be losing their concentration.

Nervous system: It causes pain, ringing in the ears, feeling of tiredness, thereby effecting the functioning of human system.

Sleeplessness: It affects the sleeping there by inducing the people to become restless and loose concentration and presence of mind during their activities.

Damage to material: The buildings and materials may get damaged by exposure to infrasonic/ultrasonic waves and even get collapsed.

4.4.3 Noise Control and Health Protection

A healthy human ear responds to a very wide range of SPL from the threshold of hearing at zero dB, uncomfortable at 100–120 dB, to painful at 130–140 dB. Due to the various adverse impacts of noise on humans and environment, noise should be controlled. The technique or the combination of techniques to be employed for noise control depend upon the extent of the noise reduction required, nature of the equipment used and the economy aspects of the available techniques.

The various steps involved in the noise management strategy is illustrated at Fig. 4.4. Reduction in the noise exposure time or isolation of species from the sources form part of the noise control techniques besides providing personal ear protection engineered control for noise reduction at source and/or diversion in the trajectory of sound waves.

Noise control can be implemented by the use of noise standards. These standards can be met by noise control at, source, control of sound transmission, physical separation of noise source and people and other elements as follows:

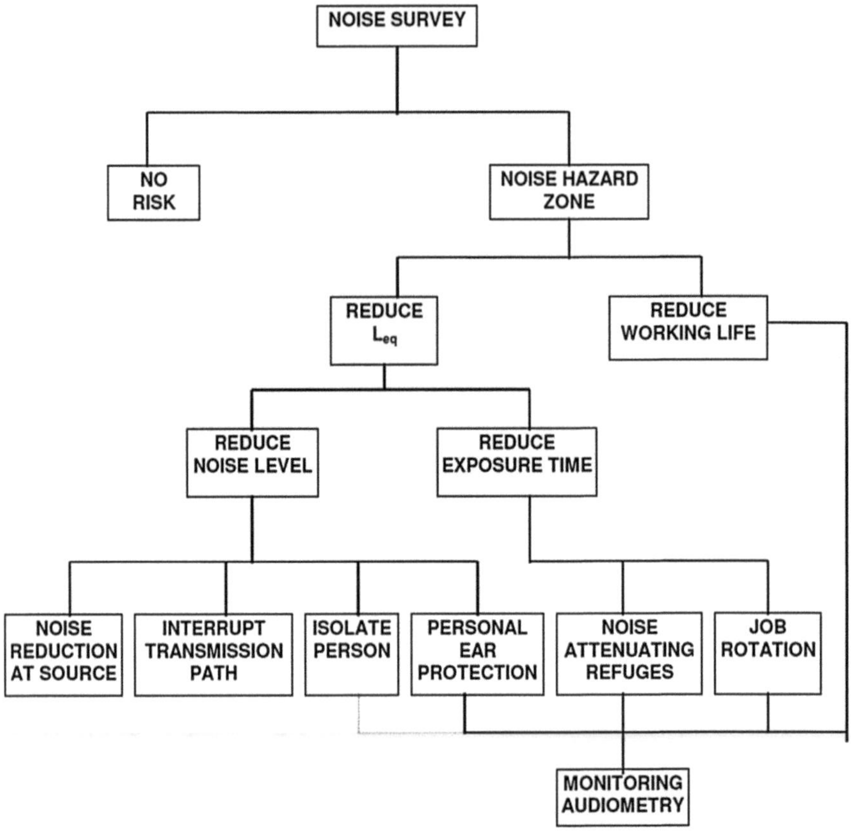

Fig. 4.4 Noise management strategy

4.4.4 Noise Control at Source

The most efficient action against excessive noise is the reduction in the noise at source. In industry, noise control technology is available for solving many typical noise problems arising from the use of machinery.

Usually, the most effective approach is to redesign or replace noisy equipment. If this is not possible, significant reductions in noise levels can be achieved by structural and mechanical modifications, or the use of mufflers, vibration isolators, and noise protection enclosures.

The noise pollution can be controlled at the source of generation itself by employing techniques like:

Reducing the noise levels from domestic sectors: The domestic noise coming from radio, tape recorders, television sets, mixers, washing machines, and cooking operations can be minimized by their selective and judicious operation. By usage of carpets or any absorbing material, the noise generated from felling of items in house can be minimized.

Maintenance of automobiles: Regular servicing and tuning of vehicles will reduce the noise levels. Fixing of silencers to automobiles, two wheelers, etc., will reduce the noise levels.

Control over vibrations: The vibrations of materials may be controlled using proper foundations, rubber padding, etc. to reduce the noise levels caused by vibrations.

Low voice speaking: Speaking at low voices enough for communication reduces the excess noise levels.

Prohibition on usage of loud speakers: By not permitting the usage of loud-speakers in the habitant zones except for important meetings/functions.

Selection of machinery: Optimum selection of machinery tools or equipment reduces excess noise levels. For example, selection of chairs, or selection of certain machinery/equipment which generate less noise (Sound) due to its superior technology, etc. is also an important factor in noise minimization strategy.

Maintenance of machines: Proper lubrication and maintenance of machines, vehicles, etc. will reduce noise levels. For example, it is a common experience that many parts of a vehicle will become loose while on a rugged path of journey. If these loose parts are not properly fitted, they will generate noise and cause annoyance to the driver/passenger. Similarly is the case of machines. Proper handling and regular maintenance is essential not only for noise control but also to improve the life of machine.

4.4.5 Control of Sound Transmission

A further reduction in noise can be obtained by increasing the distance between people and the noise source. This can be achieved in the community by planning the location of transport facilities and, in industry, by the careful selection of work sites. Sound transmission can also be controlled by the use of partitions or barriers, e.g., around particularly noisy or disturbing machinery. Reverberant noise levels can be reduced by sound-absorbing materials.

Installation of barriers between noise source and receiver can attenuate the noise levels. For a barrier to be effective, its lateral width should extend beyond the line-of-sight at least as much as the height.

The barrier may be either close to the source or receiver, subject to the condition that $R \ll D$ or in other words, to increase the traverse length for the sound wave. It should also be noted that the presence of the barrier itself can reflect sound back toward the source. At very large distances, the barrier becomes less effective because of the possibility of refractive atmospheric effects (see Fig. 4.5).

The design of the building incorporating the use of suitable noise-absorbing material for wall/door/window/ceiling will reduce the noise levels.

A sound source may be enclosed within a paneled structure such as room as a means of reducing the noise levels at the receiver. The actual difference between

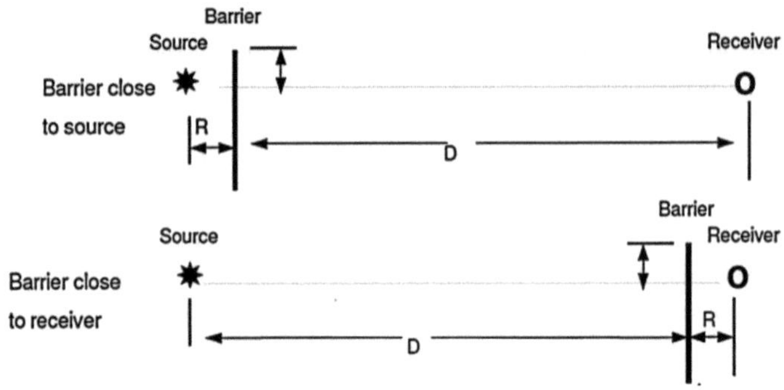

Fig. 4.5 Attenuation of the noise levels using barriers

the SPL inside and outside an enclosure depends not only on the transmission loss of the enclosure panels but also on the acoustic absorption within the enclosure and the details of the panel penetrations which may include windows or doors.

The product of *frequency of interest and surface weight of the absorbing material* is the key parameter in noise reduction through transmission loss. With conventional construction practices, the high-frequency transmission loss of a panel becomes limited to around 40 dB, owing to the transmission of sound through flanking paths other than the panel itself. Examples of such flanking are structural connections or ducts joining the two spaces on either side of the panel of interest.

4.4.6 Reduction in Length of Exposure

A reduction in the length of exposure can be used in industry to supplement the previous measures, if necessary. This may be accomplished by job rotation or by restricting the operation of the noise source.

By rotating the job between the workers working at a particular noise source or isolating a person, the adverse impacts can be reduced. Regulations prescribe that noise level of 90 dB (A) for more than 8-h continuous exposure is prohibited. Persons who are working under such conditions will be exposed to occupational health hazards. The schedule of the workers should be planned in such a way that they should not be over exposed to the high noise levels.

4.4.7 Education of Workers

It is vitally important that persons who are exposed to potentially hazardous noise levels should be educated in: (a) the possible consequences of excessive noise exposure; (b) the means of protection; and (c) the limitations of these means (e.g., improper use of earmuffs).

4.4.8 Ear Protection

If it is absolutely impossible to reduce noise to a harmless level then some form of ear protection, i.e., earplugs, earmuffs, and/or helmets, should be used. They should also be used during infrequent exposures that may not be part of a worker's normal routine.

Before employing the use of protective equipment, please recall the Fig. 4.4, wherein the various steps involved in the noise management strategy are illustrated. Protective equipment usage is the *ultimate* step in noise control technology, i.e., after noise reduction at source and/or after the diversion or engineered control of transmission path of noise.

The first step in the technique of using protective equipment is to gauge the intensity of the problem, identification of the sufferer and his exposure to the noise levels. Equipment like earmuffs, earplugs are the commonly used devices for hearing protection. Attenuation provided by earmuffs varies widely with respect to their size, shape, seal material, etc. Literature survey shows that an average noise attenuation up to 32 dB can be achieved using earmuffs.

4.5 Guidelines for Noise Control Engineer

The following documents shall always be available to the noise control engineer:

- Project approval notes;
- Basis of design;
- Project specification;
- Plot plan;
- Contractor and vendor reports on noise.

For major projects in the design and engineering phase and the procurement phase, the required documents should be submitted to the noise control engineer if close guidance by the owner is required.

The appropriate issue, or issue to be decided by the engineer concerned:

- Process flow schemes;
- Plot plan;
- Equipment summary and project specification;
- Heat exchangers (air coolers only);
- Furnaces and burners;
- Mechanical handling equipment;
- Extruders and ejectors, etc.;
- Pumps and compressors, including drivers;
- Valves (including control valves);
- Flare and vent stacks;
- External insulation and sound proofing only;
- Transformers and generators only;
- Electric motors;

- Cooling towers;
- Fired steam generators;
- Silencing equipment (silencers, enclosures, screens).

In special cases, it may be necessary also to submit the following documents:

- Process engineering flow diagram;
- Buildings.

4.6 General Noise Limits

4.6.1 Local Regulations

4.6.1.1 In-plant Noise

It shall be investigated whether any national regulations exist with respect to noise in the plant, for example, for hearing conservation, speech and work interference, accommodation.

For locations where local standards are more stringent, the applicable more stringent limits shall be stated in the project specification, or any other document defining the scope of the project.

4.6.1.2 Environmental Noise

Limits for environmental noise are not given in this book since they will depend on the local situation.

It shall be investigated whether local regulations exist with respect to environmental noise, which may include noise limits, methods of measurement and/or calculation, etc. The interpretation of such regulations shall be discussed with local authorities with the object of arriving at agreed environmental noise limits.

Environmental noise limits may be different for different times of day or night and for workdays or weekends. The most stringent of the above requirements shall be the basis of design, taking due account of the period of operation of the plant.

It shall be ensured that any allowances for occasional higher noise levels that may be acceptable to local authorities are included in the environmental noise limits, e.g., for emergencies.

Where local regulations for environmental noise do not exist, this aspect of plant design shall still be considered at the project definition stage to anticipate adverse community reactions at some later date. BS 4142 may be used for guidance.

Authorities usually specify environmental noise limits in terms of maximum allowable sound pressure levels at specified locations in the vicinity of the plant or at the plant boundary line. Such limits shall be converted into limits in terms of a maximum allowable sound power level for the plant or composing parts of the plant under consideration. The resultant limits shall be included in the project specification or any other document defining the scope of the project.

Conversion of environmental SPL into plant sound power levels and vice versa shall be carried out in accordance with EEMUA 140 (using either minimal or significant screening curves) or in accordance with a local standard agreed by the company.

4.6.2 Hearing Conservation (Work Area Noise)

The "Work Area" is defined as any position not less than 1 m from equipment surfaces accessible to personnel, or any position where a worker's ear may be exposed to noise in the normal course of his duty. It includes any platform, walkway, or ladder.

4.6.2.1 Absolute Limit

The sound pressure level anywhere in the work area shall not exceed 115 dB(A) in any situation, including emergencies such as blowing of safety/relief valves.

4.6.2.2 Work Area Limit

The sound pressure level in the work area shall not exceed 85 dB(A).

4.6.2.3 Restricted Area/Restricted Area Limit

Restricted areas are those work areas in the plant where it is not reasonably practicable to reduce the noise level below the work area limit. The absolute limit of 115 dB(A) remains valid in such areas.

If it is unavoidable that the work area limit will be exceeded around particular equipment, action shall be taken to limit the area involved as far as possible; this may include the erection of an acoustic enclosure. It is accepted that areas inside acoustic enclosures around such equipment are restricted areas.

A restricted area limit may be between 85 and 115 dB(A). However, attempts shall be made to reduce the level below 90 dB(A).

Permanent warning signs to indicate the mandatory use of ear protectors shall be erected at the boundaries of restricted areas. The signs should be of the type given in Fig. 4.6.

4.6.3 Speech and Work Interference

The following noise limits in Table 4.3 shall apply in order to reduce speech and work interference from equipment to acceptable proportions.

4.6.4 Accommodation

The sound pressure level shall not exceed 40 dB(A) inside personnel accommodation such as bedrooms and private cabins.

Fig. 4.6 Permanent warning signs to indicate the mandatory use of ear protectors

Table 4.3 Maximum allowable sound pressure level

Area description	Maximum allowable sound pressure level dB(A)
Areas in workshops and machinery buildings where communication is required;	70
Workshops for light maintenance	
Workshop offices;	60
Control rooms, not continuously manned;	
Computer rooms	
Control rooms, continuously manned	50
Open plan offices;	
Social rooms; changing rooms; wash places, and toilets	
Offices and conference rooms	45

4.6.5 Additional Restrictions for Narrow-Band or Impulsive Noise

Further restrictions shall apply if the noise contains narrow-band or impulsive components as defined in EEMUA 140 and this shall be taken into account when specifying equipment noise limits.

For environmental noise, any narrow-band or impulsive component shall be sufficiently below the broad-band noise from the plant so that it is no longer audible.

4.7 Equipment Noise Limits

Equipment noise limits shall be derived using the area noise limits and the total available plant sound power level.

4.7.1 Maximum Sound Pressure Levels for General Equipment

Where more or less (restricted area) stringent area noise limits apply, the equipment limits given below shall be adapted accordingly.

Equipment sound pressure limits shall not be exceeded anywhere at a distance of 1 m from the equipment surface.

The equipment noise limits shall be given in an overall dB(A) value or, where considered more appropriate, the corresponding octave band spectrum may be specified instead.

4.7.1.1 Equipment Emitting Continuous Noise

The maximum value for the equipment noise limit (sound pressure level) shall be 85 dB (A).

If equipment consists of components, e.g., a driver and a driven part, the above limit applied to each component separately will not ensure that the assembled equipment can give the work area limit. For such equipment, more stringent limits shall be specified on data/requisition sheets for the separate components. The permissible noise limit component shall be based on acoustic calculations. As a guidance, the following can be used:

- A reduction of 3 dB(A) shall be made for each item of a 2-component train;
- A reduction of 5 dB(A) for each item of a 3-component train.

A further reduction in the equipment noise limit shall be made when several equipment items or trains are to be mounted close together, i.e., when the distance between equipment surfaces is less than the largest equipment dimension, or when equipment are located in a reverberant area. The equipment noise limit shall be based on calculations.

4.7.1.2 Equipment Emitting Intermittent or Fluctuating Noise

Where the equipment emits an intermittent or fluctuating noise (e.g., depressuring, boiler blowdown, sump pump), the equivalent continuous sound level, Leq, over the most noisy consecutive 8-h period shall not exceed the equipment limits. The maximum level shall not be more than 10 dB(A) higher than the limit for continuous noise.

For intermittent noise, the equivalents of 85 dB(A) over 8 h are as follows in Table 4.4, provided that no significant noise [i.e., above 75 dB(A)] is emitted for the remaining time in the 8-h period.

Table 4.4 Threshold limit value (TLV)

With equipment actually operating (h)	Maximum sound pressure level with equipment operating dB(A)
8	85
4	88
2	91
1	94

4.7.1.3 Equipment Located Outside the Work Area

The maximum allowable sound pressure level at 1 m from the equipment for equipment located outside the work area. At positions which are inaccessible for personnel, such as may be the case for vent stacks and certain control valves, the allowable increase shall be:

20 * log(×) dB(A) for point sources (e.g., vent openings)

10 * log(×) dB(A) for line sources (e.g., piping)

where × is the shortest distance from the equipment under consideration to the nearest work area, expressed in meters. For valves, the distance should be taken from directly connected piping to the nearest work area.

4.7.1.4 Additional Restrictions for Narrow-Band or Impulsive Noise

Further restrictions shall apply when the noise of an equipment item contains narrow-band and/or impulsive components, as defined in EEMUA 140. The equipment noise limit shall be reduced by 5 dB(A) for such equipment.

4.7.2 Maximum Sound Pressure Levels for Specific Equipment

4.7.2.1 Valves for Control and Depressuring

Control valve noise shall be determined for three operating conditions, viz. minimum, normal, and maximum throughput. Noise limits shall not be exceeded for any of the three conditions.

4.7.2.2 Safety/Relief and Emergency Depressuring Valves

The noise from safety/relief valves and high-rate depressuring valves (and their piping), which blow under emergency conditions only, shall not exceed the absolute limit in any work area.

The party which sizes safety/relief and emergency depressurizing valves shall also be responsible for the calculation of their noise levels, according to a method to be approved by the company. The calculations however shall be confirmed by the supplier.

If it is not possible to remain within the absolute limit for safety/relief or emergency depressuring valves, they should be considered as follows:

(a) positioned well away from the work area, so that personnel shall not have access to their immediate vicinity. In this case, the maximum allowable sound pressure level, Lp, 1 m away from the valve and/or piping shall be calculated using the following equations:

$$Lp = 115 + 20 * \log(\times)\ \mathrm{dB(A)}\ \text{for point sources (e.g. relief valves),} \quad (4.3)$$

and

$$Lp = 115 + 10 * \log(\times)\ \mathrm{dB(A)}\ \text{for line sources (e.g. pipeline)} \quad (4.4)$$

where (×) is the shortest distance in meters between the source and the nearest work area;

(b) provided with a screen that shall deflect noise away from the nearest work area;

(Notes: 1- The above procedures will allow safety/relief valves and their piping to emit noise above the limits of 115 dB(A) at 1 m distance. The corresponding high levels of vibrational energy in the piping, which could cause acoustic fatigue, shall be taken into account in the design of the piping system.

2- The limit of 115 dB(A) may need to be reduced to meet the requirements of environmental noise. Repositioning and shielding shall then not be allowed.

3- Connected pipe work can also radiate excessive noise. Care shall be taken that these noise sources are taken into account).

(c) fitted with silencers or acoustic insulation; proposals shall be submitted to the company's authorities for approval.

4.7.2.3 Piping

Noise emitting from piping is of major importance in plant noise control and shall be subject to the same restrictions as Equipment General. Such noise usually has its origin in equipment such as a valve or compressor. The noise shall be controlled by selecting low-noise equipment (by design) or, when this is not reasonably practicable, by incorporating in-line silencers or acoustic insulation.

4.7.2.4 Flares

Elevated flares shall not exceed the work area limit at the perimeter of the sterile area (of at least 60 m from the flare base) when operating at flow rates up to 15 % of maximum flaring capacity. Ground flares shall not exceed the work area limit outside the windscreen or louver wall.

If the plant to which the flare is allocated is subject to environmental noise requirements, the application of low-noise flares shall be evaluated even if the flare is to be used for emergency conditions only.

4.7.3 Maximum Sound Power Level for Equipment

When an environmental noise limit is specified in terms of sound power level, the sound power limit for individual equipment shall be determined such that the sum of the levels of all equipment does not exceed the total limit.

The allocation to individual equipment items in an early stage of a project can best be performed using vendor data, data bases, and experience taking into account type, size, and speed of equipment.

Unless otherwise specified, equipment emitting intermittent or fluctuating noise shall not cause the environmental limit to be exceeded at any time (i.e., the equivalent noise-level concept).

4.7.4 Coherence of Noise Limits

The equipment noise limits of previous sections are sufficient only to ensure that in the completed plant, the work area limit will not be exceeded. Where other noise limits also apply, it shall be investigated whether they will be met.

For the purpose of this investigation, the sound power levels of individual equipment should be estimated.

Where actual sound power levels or SPL are known, they should be used instead. The sound power levels of individual equipment and their location thus form the basis of a calculation of noise levels in the plant. Calculations shall be carried out in accordance with EEMUA 140 or in accordance with a local standard agreed by the company. Screening effects of large buildings and tanks shall be taken into account where relevant.

Noise levels inside buildings or shelters shall be calculated, taking into account both the noise from equipment in the building and the noise from outside. Calculation of the attenuation of noise from outside to within a building shall be based on standard acoustic principles.

Where an environmental noise limit is specified in terms of sound power, it will be sufficient, for this purpose, to add up the sound power levels of individual equipment in order to arrive at a total sound power level per plant or composing part, see 11.4.

If the results of the investigation indicate that one or more of the limits will be exceeded, the equipment concerned should be reconsidered and replaced by equipment emitting less noise or, if this is not reasonably practicable, be treated with noise control measures such as insulation or acoustic enclosure.

The extent of noise control measures to be taken shall be considered against the severity of the requirements.

4.7.5 Data/Requisition Sheets

Equipment noise limitation sheets shall be prepared for all relevant items of equipment or an equipment train, if such a train will be provided by a single supplier, specifying the limits as given in above-mentioned sections. If the components of a train will be provided by different suppliers, separate equipment noise limitation sheets shall be prepared.

It shall be ensured that the supplier is made aware of the obligation to state noise guarantees for the equipment to be provided, for any of the conditions of operation for which the equipment may be expected to be used.

The equipment noise limitation sheet shall be returned with the tender, with its guarantee section completed by the supplier to give the following information:

(a) Unsilenced sound pressure and sound power dB(A) levels in octave bands and overall value of the equipment.
(b) Sound pressure and sound power dB(A) levels in octave bands and overall value of the equipment together with details of any silencing measures that may be necessary to meet the specified noise limits.
(c) If the specified limits cannot be met, the minimum attainable sound pressure and sound power dB(A) levels in octave bands and overall value shall be given.

Additionally, the following information shall be given where applicable:

(d) Completed data/requisition sheets of any silencers and/or acoustic enclosures, if the tender comprises such silencing equipment.

4.7.6 Equipment Selection

It shall be ensured that suppliers have provided in their tenders all the information required in accordance with above-mentioned section.

Bid comparison shall be made including the cost of all the options required to meet the specified noise limits. Where a choice can be made of equipment of low noise emission by design and equipment with "path treatment," the estimated extra cost in operation and maintenance that may result from path treatment should be capitalized and taken into account in the ultimate selection. Generally, preference shall be given to equipment of low noise emission by design. For equipment emitting fluctuating or intermittent noise, temporary excursions above the equipment noise limit may be permissible. If such excursions can be avoided by selecting a less noisy and otherwise acceptable alternative, then this shall be preferred. Where a higher extra expenditure would seem justified, the company shall be consulted.

4.8 Silencing Equipment

Silencers shall comply with appropriate codes and standards for mechanical design. The design of silencers shall ensure that any internal parts that may inadvertently break loose do not partly or completely block the silencer outlet or damage downstream equipment (e.g., compressors).

Acoustic enclosures may be either of the "close fitting" or "walk-in" (large) type. Large enclosures shall be designed in such a way that operating/maintenance personnel can conduct their work without being hampered in their movements.

Large enclosures shall also be subdivided in such a way that personnel servicing the equipment during partial shutdown are not exposed to excessive noise generated by other equipment that may also be in the enclosure. This can be achieved for example by a separating wall between two independent trains of equipment.

Silencers and/or acoustic enclosures may be included in the tenders for noisy process equipment. When the supplier has been selected, details of such silencing equipment shall be agreed upon between the supplier and the company. The company may choose to order silencing equipment separately from the process equipment.

Noise emission from certain silencing equipment (e.g., silencers on vent stacks, acoustic insulation on pipes) is not always the responsibility of any particular equipment supplied. It shall be ensured that such silencing equipment is incorporated in the design and that the relevant data/requisition sheets are prepared.

It shall be ensured that sound-absorbing materials or constructions are incorporated in the design of potentially reverberant spaces. The minimum absorption coefficient, averaged over walls and ceiling, shall be 0.2 for frequencies above 200 Hz.

For some equipment, calculations or measurements may indicate that the noise limits will be exceeded, but with a margin of uncertainty extending down to levels below the limit (i.e., there is a possibility that the limits may be met). For such equipment, the application of noise control measures shall not be postponed until actual noise levels have been obtained after start-up, unless specifically agreed with the company.

Silencers to suppress commissioning noise only may be of a movable type. Silencers to reduce noise in transient operations such as start-up, shutdown, shall be permanent.

4.9 Project Noise Documents

4.9.1 Information for the Noise Control Authorities

For the provision of the required information within the company's office, a list of documents that should be made available is as follows:

- Basis of design,
- Project specification,
- Plot plan and area classification,
- Noise allocation report (if applicable),
- Noise report,
- Noise verification study.

For major projects in the design and engineering phase and the procurement phase, the following documents should be submitted to the noise control authorities if close guidance by the company is required. The appropriate issue or issues shall be decided by the authorities concerned.

- Equipment summary and project specification.
- Equipment data requisitions for:
  - Heat exchangers (air coolers only),
  - Furnaces, reformes and burners,
  - Mechanical handling equipment,
  - Extruders and ejectors,
  - Pumps and compressors, including drives,
  - Valves,
  - Flare and vent stacks,
  - External insulation and sound proofing only,
  - T and generators,
  - Electric motors,
  - Cooling towers,
  - Fired steam generators,
  - Silencing equipment (silencers, enclosures, and screens).
- Data on the acoustic properties of buildings (in special cases only).

4.9.2 Detailed Engineering Phase Reports

Where environmental noise limits or a plant sound power limit are specified, a report shall be prepared to show how sound power levels are allocated to various equipment items. The report, with figures based on manufacturer's information, experience, or data bases, serves to set equipment noise limits and to anticipate noise abatement measures. The report shall be submitted to the company for approval before data/requisition sheets are released for tendering.

When a preliminary noise allocation report has been prepared by the relevant company's authorities as part of the project definition document, a reissue shall be submitted incorporating the contractor's assessment.

In any event, a report shall be prepared for approval by the company after the major part of the equipment selection has taken place. The report shall give sufficient data and calculations to demonstrate to the satisfaction of the company that the acoustic design of the plant will meet the requirements.

4.10 Acoustic Design

This section deals with acoustic aspect and is intended for noise control engineer, responsible for the acoustic designs of the plant.

4.10.1 Acoustical Insulation for Pipes, Valves, and Flanges

The noise radiated by the wall of the pipe is usually generated by equipment connected to the pipe, such as compressor, pumps, valves, or ejectors. These noise sources may cause long section of pipe to radiate noise because noise will travel in the pipe with little reduction. The noise radiation may be reduced by acoustic insulation.

4.10.2 General Requirements for Vent Silencers

These requirements shall apply to vent silencers for steam, hydrocarbon vapors, and other gases. It does not apply to vents used for particulate or polymerizing materials where special designs may be required to prevent clogging.

The silencer shall be constructed of materials compatible with the fluid being vented and all steel surfaces shall have adequate weather protection. The manufacturer shall provide details of all materials used and of the weather protection applied.

4.10.3 Acoustical Barrier and Enclosures

The most fundamental approach to noise reduction is to a wall between the sound source and receiver. The wall can take the form of:

1. An enclosure of the noise source (machinery),
2. The enclosure of the receiver (employee),
3. A barrier between the two.

4.10.3.1 Machine Enclosures

The isolation of noise-producing machinery by means of acoustical enclosures provides the greatest noise reduction. Noise reductions of 20–30 dBA are common with machine enclosures, and with special isolation treatment, noise reductions above 50 dBA can be achieved. Following factors are the main aspects of engineering enclosure design which should be considered:

1. Design guidelines for acoustical effectiveness,
2. Consideration of machine operational requirements,
3. Insuring production compatibility,
4. Maintaining employee safety and welfare.

4.10.3.2 Employee Enclosure Design

In designing employee enclosure, the following important factors should be considered:

1. Location,
2. Size,
3. Visibility,
4. Proximity.

4.10.3.3 Enclosure Types

Three design approaches should be considered for machine enclosures, depending upon machine operational requirements and the extent of noise reduction required. These are as follows:

1. Localized enclosure,
2. Partial enclosure,
3. Complete enclosure.

4.10.3.4 Localized Enclosures

In many machines, high noise levels are associated with only localized machine elements. Where isolated noise sources are identified on a machine, it is more feasible to enclose only a small area, rather than the entire machine.

As a general guideline, acoustical enclosure panels should never be attached directly to machine elements which have high vibration levels. Where panels are machine mounted, vibration isolation techniques should be employed as shown in Fig. 4.7. To insure minimum sound radiation due to panel vibration, damping treatment shall also be applied to machine-mounted enclosures.

4.10.3.5 Partial Enclosures

A machine enclosure which has at least one open side or a very large opening should be considered as a partial enclosure.

A partial machine enclosure will provide virtually no noise reduction for an employee working directly at the machine unless it:

(a) Interrupts the line-of-sight path between the machine and the operator's hearing zone, or

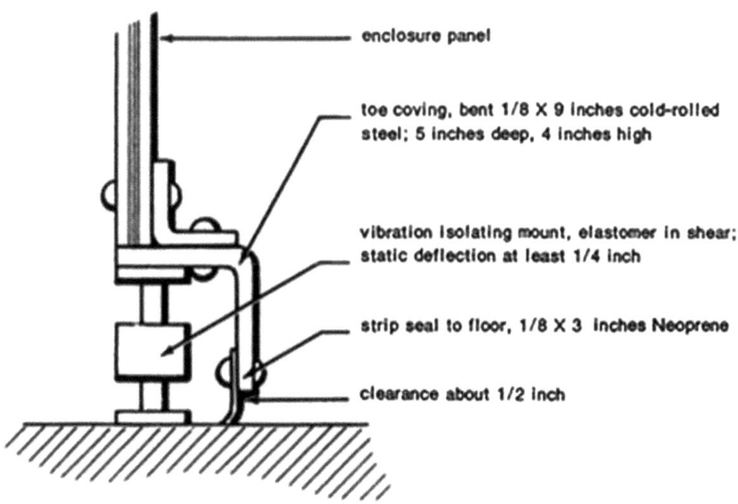

Fig. 4.7 Vibration isolation and toe covering

(b) Absorbs sound which is reflected from the machine to the operator by an adjacent wall or ceiling.

4.10.3.6 Opening in an Enclosure

For effectiveness of an acoustical enclosure, openings should be prevented or minimized. Where openings are necessary, such as for ventilation, silencers or "sound traps" can be installed.

4.10.3.7 Vibration Isolation

Vibration is transmitted through the floor to enclosures. The large surface area enclosure panels will become a noise source. This problem can be avoided by the mounting of the machine on vibration isolators. The typical vibration levels of a bolted and isolated blanking press, shown in Table 4.5, indicate the effectiveness of vibration isolation.

4.10.3.8 Non-acoustical Requirements

In addition to the design of an acoustical enclosure to meet noise reduction objectives, the following additional design requirements should be considered:

1. The enclosure shall be properly ventilated to prevent heat buildup.
2. Operational accessibility shall be provided to meet production requirements.
3. Localized and complete accessibility shall be provided for maintenance.
4. If audible signals are utilized to assess machine performance, alternate detection systems shall be installed.
5. Supply systems must be provided to the enclosed machinery to meet energy and process requirements.
6. In-feed and out-feed openings shall be designed which provide noise attenuation consistent with the total enclosure system but which will not obstruct material flow.
7. An internal lighting system shall be installed if the enclosure shadows exterior lighting excessively.
8. Protection shall be provided against employee and vehicular (lift trucks, etc.) damage.
9. Protection shall be provided against operational abuse: moisture, water spray, oil, grease, dirt, erosion by fluid flow, corrosive air, etc.

Table 4.5 Linear vibration levels in decibelS RE 1.0 g for 250-ton blanking press

Structural measurement location	Press bolted	Press isolated
Press leg	+22	+7
Foundation	−3	−30
Floor	−21	−41
Building column	−10	−25

10. Flame-spread and fire-endurance limits shall be specified for all materials. Firebreak requirements shall be employed on all ducts, pipe runs, and shafts. Smoke or temperature alarms may also be considered for enclosures.

4.11 Vibration Control

Acoustical radiation from a vibrating surface is one of the two methods by which noise will radiate. In this part, the radiation of sound from vibration, and the control of vibrational energy through isolation and damping will be discussed.

4.11.1 Vibration Radiation

Sound is produced by the movement of a vibrating structure which, in turn, sets into motion the air molecules which are coupled to it.
For any mechanical system, vibration and noise must be considered as follows:

(a) Vibration energy and
(b) System response.

System response is dependent upon the following two facts:

1. Vibration response:
   (a) Mass;
   (b) Stiffness;
   (c) Damping.
2. Acoustic radiation efficiency:
   (a) Surface area;
   (b) Critical coincidence frequency.

4.11.2 Vibration Isolation

To control vibrationally induced noise is to isolate the driving force or the machinery from the adjacent structure. Machinery which has a driving force frequency in the 10–60 Hz range and is rigidly mounted to a building foundation will create a very annoying environment for plant and office employees. Vibration isolators will generally reduce this annoyance.

4.11.3 General Vibration Considerations

1. Always specify that rotating machinery be statically and dynamically balanced.
2. Design-supporting structures to have no natural frequencies within 30 % of the operating speed of the machinery.
3. Choose equipment which minimizes vibrational problems. (if a compressor must be located in a critical area, a quieter centrifugal type can be used instead of the noisier reciprocating one).

4. Use vibration isolating mounts as required on large fans, transformers, etc. Use flexible connections to isolated pipes and ductwork.
5. Floor slabs should be heavier and stiffer to support machinery that is a potential vibration source.
6. Weight of accelerated parts should be reduced whenever possible.
7. All moving parts should have smooth finishes.

4.11.4 Vibration Damping

The physical mechanism of most noise problems is acoustical radiation from vibrating structures, and it reduces the noise by means of vibration damping. The term "damping" refers to the design property of materials which converts vibrational energy into heat energy. Figure 4.8 shows design curves for isolation efficiency versus frequency.

4.11.5 General Equipment Considerations

Consider the following items in designing any equipment for noise reduction:
(a) Balance equipment to reduce vibration.
(b) Reduce weight of accelerated parts where possible.

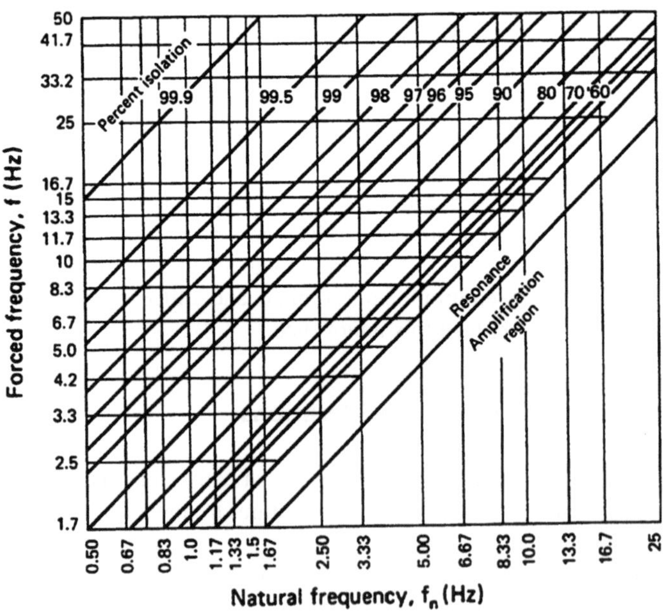

Fig. 4.8 Design curves for isolation efficiency versus frequency

(c) Reduce accelerations and decelerations of machine parts to the minimum, while still performing the required function.
(d) Check that all moving parts have smooth finishes.
(e) Reduce mechanical noise where possible.
(f) Enclose the sound source.
(g) Use acoustical absorption materials such as glass fiber to absorb sound.
(h) Reduce impact sound by using soft surfaces.
(i) Provide vibration isolators and flexible connectors.
(j) Reduce air, liquid, or gas turbulence where possible.
(k) Cover any holes in enclosures.

When the noise cannot be treated at the source, either external silencers, enclosures, barriers, or path treatment should be provided.

Glossary

Abandonment Discontinued use of a system component or components by removing them or rendering them inaccessible and inoperable.

Abatement (1) The act or process of reducing the intensity of pollution. (2) The use of some method of abating pollution. (3) Putting an end to an undesirable or unlawful condition affecting the wastewater collection system. A property owner found to have inflow sources connected to the collection system may be issued a "Notice of Abatement." Such notices will usually describe the violation, suggest corrective measures, and grant a period of time for compliance.

Absorption Process of incorporation or assimilation by which one substance is physically taken into and included with another substance, for example, bacteria assimilating nutrients from effluent. The penetration of atoms, ions, or molecules into the bulk mass of a substance. A physico-chemical process in which a substance associates with another to form a homogeneous mixture presenting the characteristics of a solution.

Abstraction The removal of water from any source, either permanently or temporarily, so that it ceases to be part of the resources of that area or is transferred to another source within the area.

Access hole A hole in the duct, at the extremity of a sampling line, through which sampling is undertaken.

Acid rain Air pollution is produced when acid chemicals are incorporated into rain, snow, fog, or mist. The "acid" in acid rain comes from sulfur oxides and nitrogen oxides, from the products of burning coal and other fuels, and from certain industrial processes. The sulfur oxides and nitrogen oxides are related to two strong acids: sulfuric acid and nitric acid. When sulfur dioxide and nitrogen oxides are released from power plants and other sources, winds blow them far from their source. If the acid chemicals in the air are blown into areas where the weather is wet, the acids can fall to Earth in the rain, snow, fog, or mist. In areas where the weather is dry, the acid chemicals may become incorporated into dusts or smokes. Acid rain can damage the environment, human health, and property.

A. Bahadori, *Pollution Control in Oil, Gas and Chemical Plants*,
DOI: 10.1007/978-3-319-01234-6,

Activated carbon Treatment A process intended for the removal of dissolved and colloidal organic substances from water and wastewater by absorption on activated carbon, for example, for the amelioration of taste, odor, or color.

Activated sludge A flocculent microbial/mass, produced when sewage is continuously aerated. Sludge that has undergone flocculation forming a bacterial culture typically carried out in tanks. Can be extended with aeration.

Activated sludge process Wastewater treatment process that uses activated sludge to biologically convert non-settleable (suspended, dissolved, and colloidal) organic materials to a settleable product using aerobic and facultative microorganisms, typically followed by clarification and sludge return.

Acute A stimulus severe enough to rapidly induce an effect; in aquatic toxicity tests, an effect observed in 96 h or less is typically considered acute. When referring to aquatic toxicology or human health, an acute effect is not always measured in terms of lethality.

Acute exposure One or a series of short-term exposures, generally lasting less than 24 h.

Additive Product added to a sewage treatment system marketed to improve performance.

Adsorption Adhesion of a substance to the surface of solid bodies or liquids with which it is in contact. A physical process in which the molecules of a gas, of dissolved substances, or of liquids adhere in extremely thin layers to the exposed surface of solid substances with which they come into contact. The retention of atoms, ions, or molecules onto the surface of another substance.

Advanced primary treatment The use of special additives to raw wastewater to cause flocculation or clumping to help settling before the primary treatment such as screening.

Advanced wastewater treatment Any advanced process used above and beyond the defacto typical minimum primary and secondary wastewater treatment.

Aeration The process of method of bringing about intimate contact between air and liquid. The introduction of air into a liquid.

Aeration, active Introduction of air via either mechanical means or diffused aeration.

Aeration, diffused Process of introducing air bubbles under pressure into a treatment unit using a compressor or blower and a diffuser.

Aeration liquor Mixed liquor. The contents of the aeration tank including living organisms and material carried into the tank by either untreated wastewater or primary effluent.

Aeration, mechanical Process of introducing air into a treatment component by physical agitation using a device such as a paddle, paddle wheel, spray nozzle, or turbine.

Aeration, passive Process of introducing air into a treatment component without mechanical means.

Aeration tank A chamber for injecting air into water. The tank where raw or settled wastewater is mixed with return sludge and aerated. The same as "aeration bay," "aerator," or "reactor."

Aerobic In the presence of, or requiring, oxygen. Wastewater treatment depending on oxygen for bacterial breakdown of waste.

Aerobic bacteria Bacteria that require free (elementary) oxygen for growth.

Aerobic condition Descriptive of a condition in which dissolved oxygen is present.

Aerobic treatment unit (ATU) A mechanical wastewater treatment unit that provides secondary wastewater treatment for a single home, a cluster of homes, or a commercial establishment by mixing air (oxygen) and aerobic and facultative microbes with the wastewater. ATUs typically use a suspended growth process (such as activated sludge-extended aeration and batch reactors), a fixed-film process (similar to a trickling filter), or a combination of the two treatment processes.

Aerosol A suspension, in a gaseous medium, of solid particles, liquid particles, or solid particles and liquid particles, having a negligible falling velocity. (Note: In physics, an upper value is arbitrarily assigned to the dimensions of particles capable of constituting an aerosol, adopting for the place being considered a maximum limit for the falling velocity. It is defined as being that of a spherical particle of density equal to 103 kg/m^3 and of diameter 100 μm, falling under the effect of its own weight in an immobile gas at a temperature of 20 °C and at a pressure of 101.3 kPa. In air, under a gravitational acceleration of 9,81 m/s^2, this speed is 0.25 m/s.)

Aeration chamber Chamber or tank in which wastewater is brought into contact with air to facilitate biological degradation such as in (but not limited to) the activated sludge process.

Aeration system Piping, diffusers, air source, vents, and all other necessary devices for an active aeration process.

Aerobic decomposition Decomposition and decay of organic material in the presence of "free" or dissolved oxygen.

Aerobic digestion The breakdown of wastes by microorganisms in the presence of dissolved oxygen. This digestion process may be used to treat only waste activated sludge, or trickling filter sludge and primary (raw) sludge, or waste sludge from activated sludge treatment plants designed without primary settling. The sludge to be treated is placed in a large aerated tank where aerobic microorganisms decompose the organic matter in the sludge. This is an extension of the activated sludge process.

Aerobic process A waste treatment process conducted under aerobic (in the presence of "free" or dissolved oxygen) conditions.

Aerobic treatment unit (ATU) 1. Treatment component that utilizes oxygen to degrade or decompose wastewater, with or without mechanical means; 2. term traditionally used to describe proprietary devices that use direct introduction of air into wastewater by mechanical means to maintain aerobic conditions within the pretreatment component.

Aerobic wastewater treatment Oxygen-dependent wastewater treatment requiring the presence of oxygen for aerobic bacterial breakdown of waste.

Agglomerate A collection of solid particles adhering to each other.

Agglomeration The grouping, or coming together of dispersed suspended matter into larger particles, called "iloc," which settle more rapidly. The action leading to the formation of agglomerates. The coalescence of flocks or particles of suspended matter to form larger flocks or particles which settle or may be caused to float more readily.

Agglutination The action of joining, by impact, solid particles coated with a thin adhesive layer or of trapping solid particles by impact on a surface coated with adhesive.

Aggregate 1. Primary soil particles that cohere to each other more strongly than other surrounding particles; 2. Naturally occurring inorganic material (crushed rock or gravel) screened to sizes for various uses; 3. A relatively stable assembly of dry particles, formed under the influence of physical forces.

Air pollutant Any material emitted into the atmosphere either by human activity or by natural processes and adversely affecting man or the environment.

Air pollution Usually, the presence of substances in the atmosphere resulting from either human activity or natural processes, present in sufficient concentration, for a sufficient time and under circumstances such as to interfere with comfort, health, or welfare of persons or the environment.

Air quality index (AQI) A numerical index used for reporting severity of air pollution levels to the public. It replaces the formerly used pollutant standards index (PSI). AQI incorporates five criteria pollutants—ozone, particulate matter, carbon monoxide, sulfur dioxide, and nitrogen dioxide—into a single index. AQI levels range from 0 (good air quality) to 500 (hazardous air quality). The higher the index, the higher the level of pollutants and the greater the likelihood of health effects.

Air toxic Any air pollutant for which a national ambient air quality standard does not exist (i.e., excluding ozone, carbon monoxide, PM10, sulfur dioxide, and nitrogen dioxide) that may reasonably be anticipated to cause cancer, developmental effects, reproductive dysfunctions, neurological disorders, heritable gene mutations or other serious or irreversible chronic or acute health effects in humans.

Algal bloom A large, visible mass of algae found in water bodies such as lakes or estuaries. Blooms occur most often during warm weather, but may also occur at other times of the year. Color ranges from green to red.

Algae Microscopic plants which contain chlorophyll and live floating or suspended in water. They also may be attached to structures, rocks, or other similar substances. Algae produce oxygen during sunlight hours and use oxygen during the night hours. Their biological activities appreciably affect the pH and dissolved oxygen of the water.

Alkaline The condition of water or soil that contains a sufficient amount of alkali substances to raise the pH above 7.0.

Alkalinity A measure of a substances ability to neutralize acid. Water containing carbonates, bicarbonates, hydroxides, and occasionally borates, silicates, and phosphates can be alkaline. Alkaline substances have a pH value over 7.

Alpha factor In an activated sludge plant, the ratio of the oxygen transfer coefficient in mixed liquor to the oxygen transfer coefficient in clean water.

Alternative fuels Fuels that can replace ordinary gasoline. Alternative fuels may have particularly desirable energy efficiency and pollution reduction features. Alternative fuels include compressed natural gas, alcohols, liquefied petroleum gas (LPG), and electricity.

Ambient Pertaining to the current environmental condition; ambient monitoring evaluates water quality periodically at fixed locations; data collected over long periods of time help determine the status and trends of water quality of a particular water body.

Ammonia (NH_3) A chemical combination of hydrogen (H) and nitrogen (N) found extensively in nature. An indicator of fresh pollution.

Ammonia stripping A method of removing ammonia content from water by making it alkaline and aerating.

Anaerobic Wastewater treatment in which bacteria break down waste without using oxygen.

Anaerobic bacteria Bacteria that grow in the absence of free oxygen and derive oxygen from breaking down complex substances.

Anaerobic wastewater treatment Wastewater treatment in the absence of oxygen; anaerobic bacteria break down waste.

Anoxic Condition in which all constituents are in their reduced form (no oxidants present); conditions in a septic tank are generally anaerobic, but not anoxic.

Antidegradation Policies which ensure protection of water quality for a particular water body where the water quality exceeds levels necessary to protect fish and wildlife propagation and recreation on and in the water. This also includes special protection of waters designated as outstanding natural resource waters. Antidegradation plans are adopted by each state to minimize adverse effects on water.

An open dump An open dump is a facility that does not meet the criteria for a sanitary landfill and is not a facility for the disposal of hazardous waste.

Aquatic bench A 10–15-foot-wide bench around the inside perimeter of a permanent pool that ranges in depth from 0–12 inches—vegetated with emergent plants, it augments pollutant removal, provides habitats, protects the shoreline from the effects of water level fluctuations, and enhances safety.

Aquifer A geologic formation capable of transmitting significant quantities of groundwater under normal hydraulic gradients. Water-bearing formation (bed or stratum) of permeable rock, sand, or gravel capable of yielding significant quantities of water.An underground geological formation or group of formations containing usable amounts of groundwater that can supply wells and springs.

Aquitard A geologic formation that may contain groundwater but not capable of transmitting significant quantities of groundwater under normal hydraulic gradients.

Area sources Those sources for which a methodology is used to estimate emissions. This can include area-wide, mobile, and natural sources, and also groups of stationary sources (such as dry cleaners and gas stations). The federal air toxic program defines a source that emits less than 10 tons per year of a single hazardous air pollutant (HAP) or 25 tons per year of all HAPs as an area source.

Areal fill Above-grade soil treatment area designed and installed such that the entire infiltrative surface is located above the original ground elevation using suitable imported soil material for fill; utilizes gravity, pressure-dosed gravity, or low-pressure distribution; a final cover of suitable soil stabilizes the completed installation and supports vegetative growth.

Aromatic Aromatic is of or relating to organic compounds that resemble benzene in chemical behavior.

Ash The solid residue of effectively complete combustion.

Assessment An evaluation of watersheds based on the presence or lack of specific non-point source indicators.

Assimilative capacity The amount of pollutants that a water body may absorb while maintaining corresponding water quality standards, including protection of aquatic life and human health.

Atmospheric deposition Process by which atmospheric pollutants reach the land surface either as dry deposition or as dissolved or particulate matter contained in precipitation.

Attached-growth process Configuration wherein the microorganisms responsible for treatment colonize a fixed medium.

Attainment area A geographic area in which levels of a criteria air pollutant meet the health-based primary standard (national ambient air quality standard, or NAAQS) for the pollutant. An area may have an acceptable level for one criteria air pollutant, but may have unacceptable levels for others. Thus, an area could be both attainment and non-attainment at the same time.

Attenuation Attenuation is to reduce or lessen in amount (e.g., a reduction in the amount of contaminants in a plume as it migrates from the source).

Average land cover condition Percentage of impervious cover considered to generate an equivalent amount of phosphorus as the total combined land uses within the Chesapeake Bay watershed at the time of the CBA adoption, assumed to be 16 %.

Average monthly discharge limitations The highest allowable average of daily discharges over a calendar month, calculated as the sum of all daily discharges measured during that month divided by the number of days on which monitoring was performed (except in the case of fecal coliform).

Average weekly discharge limitation The highest allowable average of daily discharges over a calendar week, calculated as the sum of all daily discharges measured during a calendar week divided by the number of daily discharges measured during that week.

A-weighted scale It resembles the audible response of human ear. Represented as dB(A).

Backfill 1. Material placed in an excavation; 2. to place material in an excavation; 3. portion of an excavation above the haunch zone; for straight-walled tanks or structures, that portion of an excavation above the bedding.

Backfill, initial Portion of an excavation above the haunch zone or bedding with a depth of 6–12 inches (15–30 cm) above the piping, conduit tank, or structure.

Backfill final Portion of an excavation extending from above the initial backfill to final grade.

Backflow Reverse direction of flow, with liquid returning to the source.

Backflow prevention device Any device, method, or configuration used to prevent a reversal of flow.

Backflush To reverse the direction of flow to clean laterals or filter media.

Background level Amount of a substance expected to occur naturally in the environment.

Backwash Cleaning a granular media bed by means of an upward flow of clean water.

Backwashing The operation of cleaning a filter with water, or with air and water, by reversing the direction of flow.

Bacteria Single-cell microscopic living organisms lacking chlorophyll, which digest many organic and inorganic substances. An essential part of the ecosystem including within human beings. Microscopic one-celled organisms which live everywhere and perform a variety of functions. While decomposing organic matter in water, bacteria can greatly reduce the amount of oxygen in the water.

Bacteria, aerobic Bacteria that can metabolize only in the presence of molecular oxygen.

Bacteria, anaerobic Bacteria that is able to metabolize in the absence of molecular oxygen.

Bacteria, facultative Bacteria that can metabolize with or without molecular oxygen present in the environment.

Bacteria, mesophilic Bacteria which grow best at temperatures between 20 and 50 °C (68 and 122 °F) with optimum growth between 25 and 40 °C (77 and 104 °F).

Bacteria, psychrophilic Bacteria which grow best at temperatures between 10 and 30 °C (50 and 86 °F) with optimum growth between 12 and 18 °C (54 and 64 ° F).

Bacteria, thermophilic Bacteria which grow best at temperatures between 35 and 75 °C (95 and 167 °F) with optimum growth between 55 and 65 °C (131 and 149 °F).

Baffle Physical barrier placed in a component to dissipate energy, direct flow, retain solids and FOG, and/or draw water from a specific depth.

Ballasting (Deballasting) The act of taking on (discharging) ballast water.

Baseflow The amount of water in a stream that results from groundwater discharge.

Beneficial use Use of a (water) resource that includes, but is not limited to, domestic (including public water supply), agricultural, commercial, industrial, water-based recreational uses, and the propagation and growth of aquatic life.

Benthic Pertains to the bottom, or bed, of a water body.

Best available technology economically achievable Technology-based standard established by the clean water act (CWA) as the most appropriate means available on a national basis for controlling the direct discharge of toxic and non-conventional pollutants to navigable waters. BAT effluent limitation guidelines, in general, represent the best existing performance of treatment technologies that are economically achievable within an industrial point source category or subcategory.

Best management practice (BMP) A structural or non-structural method, activity, maintenance procedure, or other management practices used singularly or in combination to reduce non-point source inputs to receiving waters in order to achieve water quality protection goals. Examples include animal waste management systems, conservation tillage systems, vegetated filter strips.

Best practical technology currently available (BPTCA) BPTCA is generally defined as the equivalent of secondary treatment presently being practiced in the particular industrial subcategory. For example, this may be considered as the activated sludge process in the organic chemical industry.

Best usage The most appropriate uses of a water body as designated by the Environmental Management Commission given the characteristics of the water body and surrounding area. Best uses may include the use of public water supplies; protection and propagation of fish, shellfish, and wildlife; recreation in and on the water; as well as uses for agriculture, industry, and navigation.

Bioassay A test used to evaluate the relative potency of a chemical or a mixture of chemicals by comparing its effect on a living organism with the effect of a standard preparation on the same type of organism.

Biochemical oxygen demand (BOD) The rate at which organisms use the oxygen in water or wastewater while stabilizing decomposable organic matter under aerobic conditions. In decomposition, organic matter serves as food for the bacteria and energy results from its oxidation. BOD measurements are used as a measure of the organic strength of wastes in water. Biochemical oxygen demand is oxygen demand by microorganisms during stabilization of organic matter under prescribed condition, usually over a 5-day period; BOD5 specifically denotes the oxygen demand over a 5-day period at 20 °C. The BOD water test is used to determine how much oxygen is being used by aerobic microorganism in the water to decompose organic matter, the amount of oxygen required by aerobic biological processes to break down the organic matter in water. BOD is a measure of the pollutional strength of biodegradable waste on dissolved oxygen in water.

Biodegradation The breakdown of organic matter by bacteria to more stable forms which will not create a nuisance or give off foul odors. A process by which microbial organisms transform or alter through enzymatic action the structure of chemicals introduced into the environment. Molecular degradation of organic matter resulting from the complex actions of living organisms, ordinarily in an aqueous medium.

Bioengineering A low-tech construction method using living plants as a functioning, self-sustaining part of the system. Examples include control of erosion of stream banks, water quality treatment, and flood control and habitat restoration.

Biofilm Thin coating of microbial growth, organic matter, and microbial secretions on a solid substrate particle.

Biofilm (of a Sand Filter) The film, consisting of living organisms, which forms on the surface of a slow sand filter and which is considered to provide an important part of the effective filtering zone.

Biofilter Media filter in which the medium used is biological in origin (i.e., peat or coir).

Biological filter A bed of relatively inert material (such as clay, molded plastics, clinker, etc) to promote or assist natural aerobic degradation of sewage.

Biological nutrient removal (BNR) Use of microbiological activity for the removal of nitrogen and phosphorus in a wastewater treatment system.

Biological oxidation The process by which bacteria and other types of microorganisms consume dissolved oxygen and organic substances in wastewater, using the energy released to convert organic carbon into carbon dioxide and cellular material.

Biological water quality sampling The use of biological or ecological characteristics, such as the growth, survival, and reproduction of an aquatic species, the diversity, structure, and functioning of an aquatic community, and characterizations of aquatic habitat, to measure the "effects" of environmental impairment.

Biomass A mass or clump of organic material consisting of living organisms feeding on the wastes in wastewater, dead organisms, and other debris. The amount of living matter in a given area or volume. The total mass of living material in a given water body.

Bioretention basin Water quality BMP engineered to filter the water quality volume through an engineered planting bed, consisting of a vegetated surface layer (vegetation, mulch, ground cover), planting soil and sand bed (optional), and into the in situ material—also called rain gardens.

Bioretention filter Bioretention basin with the addition of a sand layer and collector pipe system beneath the planting bed.

Bio-solids Rich organic material leftover from aerobic wastewater treatment, essentially dewatered sludge that can be reused.

Biota The living components of an aquatic system including flora and fauna of a region.

Biotic index A numerical value used to describe the biota of a water body, serving to indicate its biological quality.

Blackwater Portion of the wastewater stream that originates from toilet fixtures, dishwashers, and food preparation sinks.

Bleed To drain a liquid or gas, as in bleeding accumulated air from a waterline or bleeding (draining) a trap of accumulated water.

BOD5 Refers to the five-day biochemical oxygen demand. The total amount of oxygen used by microorganisms decomposing organic matter increases each day until the ultimate BOD is reached, usually in 50 to 70 days. BOD usually refers to the five-day BOD or BOD5. The amount of dissolved oxygen consumed in five days by biological process breaking down organic matter.

Branches Branches are collection from various drain funnels, catch basins, and area drains and tie into sublaterals. They are called T, Y, T-Y, double Y, and V branches according to their respective shapes.

Btu British thermal unit, a unit of energy: 1 Btu = 1,060 J.

Bubble, coarse, Bubble of 3–8 mm diameter generated by an air diffuser.

Bubble, fine Bubble of 0.2–3 mm diameter generated by an air diffuser.

Bubble, micro Bubble of less than 0.2 mm diameter generated by an air diffuser.

Bulk density The amount of mass of a soil per unit volume of soil, where mass is measured after all water has been extracted and total volume includes the volume of the soil itself and the volume of air space between the soil grains.

Bulking Inability of sludge solids to separate from the liquid under quiescent conditions; under aerobic conditions may be associated with the growth of filamentous organisms, low DO, or high sludge loading rates; under anaerobic conditions may be associated with the attachment of gas bubbles to solids.

Building sewer Piping that conveys wastewater to the first system component or the sewer main.

Buoyancy The tendency of a body to float in water or other liquids; upward force that a fluid exerts on an object that is less dense than itself.

Bypass The intentional diversion of waste streams from any portion of a treatment (or pretreatment) facility.

Capillary fringe The zone of a porous medium above the water table within which the porous medium is saturated but is at less than atmospheric pressure. The capillary fringe is considered to be part of the vadose zone, but not of the unsaturated zone.

Capillary suction Process where water rises above the water table into the void spaces of a soil due to tension forces between the water and soil particles.

Capture The extraction of solid particles, liquid particles, or gases close to their sources.

Carbon dioxide (CO_2) A colorless, odorless gas that occurs naturally in the Earth's atmosphere. Significant quantities also are emitted into the air by fossil fuel combustion.

Carbon monoxide (CO) A colorless, odorless, poisonous gas, produced by incomplete burning of carbon-based fuels, including gasoline, oil, and wood. Carbon monoxide is also produced from incomplete combustion of many natural and synthetic products. For instance, cigarette smoke contains carbon monoxide. When carbon monoxide gets into the body, the carbon monoxide combines with chemicals in the blood and prevents the blood from bringing oxygen to cells, tissues, and organs. The body's parts need oxygen for energy, so high-level exposures to carbon monoxide can cause serious health effects. Massive exposures to CO can cause death. Symptoms of exposure to carbon monoxide can include vision problems, reduced alertness, and general reduction in mental and physical functions. Carbon monoxide exposures are especially harmful to people with heart, lung, and circulatory system diseases.

CAS registry number The chemical abstracts service Registry Number (CAS) is a numeric designation assigned by the American Chemical Society's Chemical Abstract Service and uniquely identifies a specific compound. This entry allows one to conclusively identify a material regardless of the name or naming system used.

Catch basins Catch basins are used to collect surface drainage and process wastes in individual drainage areas and to trap sediment at the point near the source.

Catchment area Catchment area is an area defined by a number of effluent streams which have a common discharge directed into a surface water drainage system, or water course. The area draining naturally to a water source or to a given point.

CEM Continuous emission monitoring (at stationary sources).

Cesspool A covered watertight tank used for receiving and storing sewage from premises which cannot be connected to a public sewer and where ground conditions prevent the use of a small sewage treatment works including a septic tank. A further stage of treating sewage.

Chamber Pre-formed manufactured distribution medium with an open-bottom configuration commonly used in soil treatment areas.

Channel The main flow of a natural or man-made waterway.

Channel stabilization The introduction of natural or man-made materials placed within a channel so as to prevent or minimize the erosion of the channel bed and/or banks.

Channelization The straightening of a stream, primarily a result of human activity.

Chemical oxygen demand (COD) COD is the equivalent amount of oxygen consumed under specified conditions in the chemical oxidation of the organic and oxidizable inorganic matter contained in a wastewater corrected for the influence of chlorides. In American practice, unless otherwise specified, the chemical oxidizing agent is hot acid dichromate. The COD test is used to estimate the amount of organic matter in a sample. It is a measurement of the oxygen equivalent of the materials present in the sample subject to oxidation by a strong chemical oxidant. **COD** is the amount of oxygen utilized in the chemical reactions that occur in water as a result of the addition of wastes. COD is a measure of the pollutional strength of chemical waste on dissolved oxygen in water.

Chemical tracer A chemical substance added to, or naturally present in water, to allow flow to be followed.

Chemical treatment A process involving the addition of chemicals to achieve a specific result.

Chemical water quality monitoring The direct, quantitative measurement of physical parameters, of the quantity or concentration of specific chemical elements or compounds, or of chemical reaction rates in aquatic substrates; the medium evaluated may be water, sediment, or biological tissues; the concept of chemical monitoring is based on measures of the possible "causes" of environmental impairment.

Combined sewer overflow (CSO) Sewer systems that combine sanitary waste and stormwater in instances of heavy rains, usually untreated; cities with older systems often have CSOs.

Chlorination The application of chlorine to water, sewage, or industrial wastes, generally for the purpose of disinfection, but frequently for accomplishing other biological or chemical wastewater treatment results.

Chlorine Term commonly used to describe a chlorine source such as sodium hypochlorite, a highly reactive chemical used as a disinfectant and oxidizing agent.

Chlorine, combined available Chlorine that has combined with ammonia in wastewater to form chloramines; although they are slow-reacting, chloramines also serve as disinfectants.

Chlorine residual Total amount of chlorine (free and combined available forms) remaining in water, sewage, or industrial wastes at the end of a specified contact period after the chlorination process.

Chlorofluorocarbons (CFCs) These chemicals and some related chemicals have been used in great quantities in industry, for refrigeration and air-conditioning, and in consumer products. CFCs and their relatives, when released into the air, rise into the stratosphere, a layer of the atmosphere high above the Earth. In the stratosphere, CFCs and their relatives take part in chemical reactions which result in the reduction in the stratospheric ozone layer, which protects the Earth's surface from harmful effects of radiation from the sun. The 1990 Clean Air Act includes provisions for reducing releases (emissions) and eliminating production and use of these ozone-destroying chemicals.

Chronic exposure Long-term exposure, usually lasting one year to lifetime.

Clarification Any process or combination of processes, the main purpose of which is to reduce the concentration of suspended matter in a liquid.

Clarifier A large circular or rectangular tank or basin in which water is held for a period of time during which the heavier suspended solids settle to the bottom. Clarifiers are also called settling basins and sedimentation basins. May also be a tank or basin in which wastewater is held for a period of time during which the heavier solids settle to the bottom and the lighter materials float to the water surface.

Clean fuels Low-pollution fuels that can replace ordinary gasoline. These are alternative fuels, including gasohol (gasoline–alcohol mixtures), natural gas, and LPG (liquefied petroleum gas).

Cleaning (after clogging) The removal of the deposit of solid or liquid particles which has produced clogging.

Cleaning factor The ratio of the quantity of pollutants entering a separator to the quantity leaving it.

Cleanout Device designed to provide access for the removal of deposited or accumulated materials, generally from a pipe.

Clear water Fraction of the wastewater stream including, but not limited to surface water, groundwater, condensate, ice machine drainage, and/or discharge from swimming pools, hot tubs, and water treatment devices.

Clogging The deposition, progressive or otherwise, of solid or liquid particles on or within a filter medium, causing the flow to be obstructed.

Clogging Capacity The particle mass that can be retained by equipment up to the point at which one of the specified operational limits is reached.

Coagulant A chemical that causes very fine particles to clump (floc) together into larger particles. This makes it easier to separate the solids from the liquids by settling, skimming, draining, or filtering.

Coagulant Aids Materials added to enhance the action of coagulants, generally by affecting the electrical balance of the particles.

Coagulation The clumping together of very fine particles into larger particles (floc) caused by the use of chemicals (coagulants). The chemicals neutralize the electrical charges of the fine particles, allowing them to come closer and form larger clumps. This clumping together makes it easier to separate the solids from the water by settling, skimming, draining, or filtering.

Coalescence The action by which liquid particles in suspension unit to form larger particles.

Cold climate limitations Cold temperatures, ice cover, plant dormancy, equipment performance, ice buildup, and reduced microbial action create design challenges for cold weather wastewater treatment.

Collection efficiency With regard to filters, dust separators, and droplet separators, the ratio of the quantity of particles retained by a separator to the quantity entering it (generally expressed as a percentage).

Colloids Very small solids (particulate or insoluble material in a finely divided form that remains dispersed in a liquid for a long time due to their small size and electrical charge).

Combined sewer Carries both sanitary sewage and stormwater runoff. Sewer systems in which the stormwater and sanitary waste are combined. A benefit is that non-point pollution flushed from the watershed during moderate rain is treated, but the system can be overwhelmed during severe storms, resulting in the untreated waste being flushed into the receiving waters as a combined sewer overflow (CSO).

Combined wastewater A mixture of storm or surface runoff and other wastewaters such as domestic or industrial wastewater.

Combustion Burning. Many important pollutants, such as sulfur dioxide, nitrogen oxides, and particulates, are combustion products, often products of the burning of fuels such as coal, oil, gas, and wood.

Comminution A mechanical treatment process which cuts large pieces of wastes into smaller pieces so they would not plug pipes or damage equipment (shredding).

Comminutor A device used to reduce the size of the solid chunks in wastewater by shredding (comminuting). The shredding action is like many scissors cutting or chopping to shreds all the large solid material in the wastewater.

Concentration The quantity of a solid, liquid, or gaseous material expressed as a proportion of another material in which it is contained in the form of a mixture, a suspension, or a solution.

Conduit Any channel intended for the conveyance of water, whether opened or closed.

Confined aquifer An aquifer under greater than atmospheric pressure, bounded above and below by relatively impermeable formations.

Confined animal feeding operation (CAFO) A lot or facility, together with any associated treatment works, where (1) animals have been, are, or will be stabled or confined and fed or maintained for a total of 45 days or more in any 12-month period and (2) crops, vegetation, forage growth, or postharvest residues are not sustained over any portion of the operation of the lot or facility (pertains to both operations that require a permit and non-permitted operations).

Confining layer A geologic formation exhibiting low permeability that inhibits the flow of water.

Connate Water Interstitial water of the same geological age as the surrounding rock or bed, often of poor quality and unfit for normal use (for example, potable purposes, industrial and agricultural use).

Consolidated soil When a soil is subjected to an increase in pressure due to loading at the ground surface, a readjustment in the soil structure occurs. The volume of space between the soil particles decreases, and the soil tends to settle or consolidate over time.

Constituent An essential part or component of a system or group: Examples are an ingredient of a chemical system, or a component of an alloy.

Constructed wetland A wetland constructed for the purpose of pollution control and waste management. The flow rate, residence time, and other factors are controlled to enhance the removal of BOD, SS, and N. A waterproof barrier is usually placed below the substrate to isolate the wastewater from the groundwater. Plants such as cattails, bulrushes, and reeds provide a dense cover and an oxygenating substrate for bacteria in the root zone.

Constructed stormwater wetlands Areas intentionally designed and created to emulate the water quality improvement function of wetlands for the primary purpose of removing pollutants from stormwater.

Control technology; control measures Equipment, processes, or actions used to reduce air pollution. The extent of pollution reduction varies among technologies and measures. In general, control technologies and measures that do the best job of reducing pollution will be required in the areas with the worst pollution. For example, the best available control technology/best available control measures (BACT, BACM) will be required in serious non-attainment areas for particulates, a criteria air pollutant. A similar high level of pollution reduction will be achieved with maximum achievable control technology (MACT) which will be required for sources releasing hazardous air pollutants.

Corrosion The gradual decomposition or destruction of a material by chemical action, often due to an electrochemical reaction. Corrosion may be caused by (1) stray current electrolysis, (2) galvanic corrosion caused by dissimilar metals, or (3) differential concentration cells. Corrosion starts at the surface of a material and moves inward.

Corrosion Inhibitors Substances that slow the rate of corrosion.

Corrosive A chemical that can cause burns to skin, eyes, or the respiratory system.

Corrosive Gases In water, dissolved oxygen reacts readily with metals at the anode of a corrosion cell, accelerating the rate of corrosion until a film of oxidation products such as rust forms. At the cathode where hydrogen gas may form a coating on the cathode and slow the corrosion rate, oxygen reacts rapidly with hydrogen gas forming water and again increases the rate of corrosion.

Conventional pollutants Pollutants typical of municipal sewage and for which municipal secondary treatment plants are typically designed; defined by Federal Regulation [40 CFR 401.16] as BOD, TSS, fecal coliform bacteria, oil and grease, and pH.

Conventional septic system A wastewater treatment system consisting of a septic tank and a typical trench or bed subsurface wastewater infiltration system.

Criteria air pollutants A group of very common air pollutants regulated by EPA on the basis of criteria (information on health and/or environmental effects of pollution). Criteria air pollutants are widely distributed all over the country. A National Ambient Air Quality Standard exists for each criteria pollutant (particulate matter, sulfur dioxide, nitrogen dioxide, ozone, carbon dioxide, and lead).

Critical success factors (CSFs) Are parameters that influence the likelihood of success of a particular method.

Cross-connection A connection between pipes which may cause the transfer of polluted water into a potable water supply with consequent hazard to public health. This term is also used to describe a legitimate connection between different distribution systems.

Culvert Man-made construction that diverts the natural flow of water.

Cyclone A dust separator or droplet separator utilizing essentially the centrifugal force derived from the motion of the gas.

Daily discharge The discharge of a pollutant measured during any 24-hour period that reasonably represents a calendar day for purposes of sampling. For pollutants with limitations expressed in units of mass, the daily discharge is calculated as the total mass of the pollutant discharged during the day. For pollutants with limitations expressed in other units of measurement (e.g., concentration), the daily discharge is calculated as the average measurement of the pollutant throughout the day.

Decantation The withdrawal of the supernatant liquor after settlement of suspended solids, or after separation from a liquid of higher density.

Detention Temporary impoundment or holding of stormwater runoff.

Detention basin A stormwater management facility that temporarily impounds runoff and discharges it through a hydraulic outlet structure to a downstream conveyance system.

Detritus In a biological context, organic particulate matter. In the context of sewage treatment practice, coarse debris denser than water but capable of being transported in moving water.

Dechlorination The removal of the free and combined chlorine residual to reduce the potentially toxic effects of chlorinated effluents.

Decibel, dB It is measurement unit of sound, represented by dB.

Degradation The conversion or breakdown of a substance to simpler compounds. For example, the degradation of organic matter to carbon dioxide and water.

De minimis In general, a level of emissions, etc., below which a particular process or activity is exempted or not regulated.

Denitrification An anoxic process that occurs when nitrite or nitrate ions are reduced to nitrogen gas and nitrogen bubbles are formed as a result of this process. The bubbles attach to the biological floc in the activated sludge process and float the floc to the surface of the secondary clarifiers. This condition is often the cause of rising sludge observed in secondary clarifiers or gravity thickeners.

Design population Design population means the minimum and maximum number of persons (resident and non-resident) to be served

Detention Time (retention time; residence time) The average period of time wastewater stays in a treatment system. Detention times vary for different types of wastewater treatment systems and can range from hours to weeks.

Dewatered sludge The sludge after it has been dewatered, also known as sludge cake.

Dewatered sludge cake The sludge after dewatering that is cake like, compressed. The lower the water content, the better for wastewater treatment purposes.

Dewatering Removing water from sludge or other solids. The process whereby wet sludge, usually conditioned by a coagulant, has its water content reduced by physical means.

Diesel engine A type of internal combustion engine that uses low-volatility petroleum fuel and fuel injectors and initiates combustion using compression ignition (as opposed to spark ignition that is used with gasoline engines).

Diffuser Part or device that injects air under pressure into wastewater (e.g., submerged porous plate, perforated pipe, or orifice).

Digestion The breaking down of sludge and other wastes biologically by micro-organisms. Results in by-products such as methane gas, carbon dioxide, sludge solids, and water. Aerobic digestion requires oxygen, anaerobic digestion the absence of oxygen.

Discarded material Discarded material is used or spent material which is not reused in any way and this is committed to final disposition.

Disinfection The use of chemicals to kill any disease-causing organisms in the polished wastewater. UV light can also be used.

Dispersion 1. Scattering and mixing; 2. Repellant action of an electric potential on fine particles in suspension in water, as in a stream carrying clay; 3. An operation as a result of which solid particles or liquid particles are distributed in a fluid. Also applied to a two-phase system in which one phase, known as the "dispersed medium," is distributed throughout the other, known as the "dispersion medium."

Disposal Disposal is the discharge, deposit, injection, dumping, spilling, leaking, or placing of any solid waste or hazardous waste into or on any land or water so that such solid waste or hazardous waste or any constituent thereof may enter the environment, be emitted into the air, or be discharged into any waters, including groundwaters. Therefore, waste deposited in a landfill, land farmed wastewater discharged to a basin or surface impoundment, or stormwater runoff diverted to a percolation or settling pond is disposal.

Disposal well Disposal well is a deep well used for the disposal of liquid wastes.

Dissolved oxygen (DO) Dissolved oxygen (DO) is the oxygen dissolved in sewage, water, or other liquids, usually expressed in milligrams per liter or percent of saturation. It is the test used in BOD determination. Oxygen dissolved in water and readily available to fish and other aquatic organisms.

Dissolved solids Solids physically suspended in sewage that cannot be removed by proper laboratory filtering.

Dose The amount of a pollutant that is absorbed. A level of exposure that is a function of a pollutant's concentration, the length of time a subject is exposed, and the amount of the pollutant that is absorbed. The concentration of the pollutant and the length of time that the subject is exposed to that pollutant determine dose.

Dosing, demand Configuration in which a specific volume of effluent is delivered to a component based on patterns of wastewater generation from the source.

Downgradient In the direction of decreasing static head.

Drain, interceptor Subsurface drain used to intercept and divert laterally moving groundwater or perched water away from the soil treatment area or other system components to an effective outlet.

Drainage Network of natural or artificial groundwater or surface water features including agricultural drain tile, cut banks, and ditches which intercept and divert surface water and/or lower groundwater.

Drains Drains are small sewer connections discharging through a sealed connection to the nearest catch basin from points such as pump bases, equipment drips, low points of floors, funnels.

Drawdown Drop in the liquid level of a tank as a result of some phase of operation. Lowering of the water table due to withdrawal of groundwater from a well.

Drawdown test Measurement of the drop in liquid level in a dosing tank measured over time to calculate dosing/delivery rate; may be expressed as a pump delivery rate (PDR) or siphon delivery rate.

Dredge To remove sediments from a stream bed to deepen or widen the channel.

Drinking water potable Water Water of a quality suitable for drinking purposes.

Drip system Drip system is a separate drain system for recovery of oil from contaminated fluids.

Droplet A liquid particle of small mass, capable of remaining in suspension in a gas. In some turbulent systems, for example clouds, its diameter can reach 200 μm.

Droplet separator An apparatus for separating liquid particles from a gas stream in which they are suspended.

Duct An enclosed structure through which gases travel from one point to another.

Dust Small solid particles conventionally taken as those particles below 75 μm in diameter which settle out under their own weight but which may remain suspended for some time….(see dust, sand, grit): a general term applied to solid particles of different dimensions and origin generally remaining suspended in a gas for a certain time.

Dust control The whole of the processes for the separation of solid particles from a gas stream in which they are suspended (by extension, also the activities involved in the construction and commissioning of a dust separator).

Dust separator An apparatus for separating solid particles from a gas stream in which they are suspended Note: Dust separators working on the following principles are given as examples:- gravity;- inertia;- centrifugal force;- electricity;- fibrous layer;- packed tower;- bubble washer;- spray washer;- Venturi scrubber.

Ecological engineering The design, management, or reconstruction of sustainable ecosystems that serve human needs such as providing clean water and food while requiring low energy inputs. Ecological engineering has enhanced our understanding of environmental problems such as wastewater treatment, wetland damage and mitigation, the effect of non-point pollution on ecosystems and ecosystem restoration.

Ecoregion An area of relatively homogeneous environmental conditions, usually defined by elevation, geology, and soil type. Examples include mountains, piedmont, coastal plain, sandhills, and slate belt.

Ecosystem A community of animals and plants and the physical environment in which they live.

Effluent Effluent is a liquid which flows out of a containing space, and/or sewage, water or other liquids, partially or completely treated, or in its natural state, as the case may be flowing out of a reservoir, basin, or treatment plant, or part thereof. **Any fluid discharged from a given source into the external environment. A general term describing any fluid discharged from a given source**. Water or wastewater discharge from a containing space such as a treatment plant, industrial process, or lagoon. Wastewater, treated or untreated, that flows out of a treatment plant, sewer, or industrial outfall. Generally refers to wastes discharged into surface waters.

Effluent limitation Effluent limitation is any restriction (including schedules of compliance) established by a governmental authority on quantities, rates and concentrations of chemical, physical, biological, and other constituents which are discharged from point sources into navigable waters, the waters of the contiguous zone, or the ocean.

Effluent screen Removable, cleanable (or disposable) device installed on the outlet piping of a septic tank for the purpose of retaining solids larger than a specific size and/or modulating effluent flow rate.

Effluent quality Physical, biological, and chemical characteristics of a liquid flowing from a component or device.

Elutriation A method of separating particles using the difference in apparent weight which may exist between the particles when they are suspended in a fluid.

Emerging contaminants Newly identified compounds or substances that have the potential to adversely affect public health or the environment and for which there is no currently published health standard.

Emission Release of pollutants into the air from a source. We say sources emit pollutants. Continuous emission monitoring systems (CEMS) are machines, for which some large sources are required to install, to make continuous measurements of pollutant release.

Emission inventory An estimate of the amount of pollutants emitted into the atmosphere from major mobile, stationary, area-wide, and natural source categories over a specific period of time such as a day or a year.

Emulsification Suspension of solids as a result of decreased surface tension due to the presence of an emulsifying agent or some substance that alters or prohibits normal microbial activity.

Emulsifying agent Agent capable of modifying the surface tension of emulsion droplets to prevent coalescence; examples are soap and other surface-active agents, certain proteins and gums, water-soluble cellulose derivatives, and polyhydric alcohol esters and ethers.

Emulsion A mixture of two or more liquids which cannot be combined; therefore, one liquid is "suspended" in the other.

Endogenous respiration Auto-oxidation by organisms in biological processes.

Enforcement The legal methods used to make polluters obey the Clean Air Act. Enforcement methods include citations of polluters for violations of the law (citations are much like traffic tickets), fines, and even jail terms. EPA and the state and local governments are responsible for enforcement of the Clean Air Act, but if they do not enforce the law, members of the public can sue EPA or the states to get action.

Environment The sum of all the external conditions that may act upon a living organism or community to influence its development or existence.

Environmental sensitivity Relative susceptibility of the natural environment to adverse impacts from an outside constituent.

Equivalent diameter The diameter of a spherical particle the size of which will give identical geometric, or optical, or electrical or aerodynamic behavior to that of the particle being examined. For sieves, the equivalent diameter is the diameter of the holes in a round-hole sieve which will pass the same proportion of material as will a specified square mesh sieve. It is dependent on the shape and size of the particles under examination.

Equivalent sound level, Leq It is the constant sound pressure level which would have produced the same total energy as the actual sound level over the given time. It is denoted as Leq.

Erosion Wearing away of rock or soil by the gradual detachment of soil or rock fragments by water, wind, ice, and other mechanical and chemical forces.

Estuary Coastal waters situated between rivers and near-shore ocean waters, where tidal action and river flow mix fresh and saltwater. Such areas include bays, sounds, mouths of rivers, salt marshes, and lagoons.

Ethanol Ethyl alcohol, a volatile alcohol containing two carbon groups (CH_3CH_2OH). For fuel use, ethanol is produced by fermentation of corn or other plant products.

Eutrophication Degradation of water quality due to enrichment by nutrients primarily nitrogen (N) and phosphorus (P), which results in excessive plant (principally algae) growth and decay. Low dissolved oxygen in the water is a common consequence.

Exceedance A measured level of an air pollutant higher than the national or state ambient air quality.

Existing construction (with failing sewage disposal systems) An existing structure where the sewage disposal system serving the structure has failed or is currently in violation of state law or regulations and requires correction.

Exposure The concentration of the pollutant in the air multiplied by the population exposed to that concentration over a specified time period.

Extended aeration An aeration system that adds aerobic sludge digestion to the activated sludge process.

Extension Alteration of a wastewater treatment system resulting in an increase in capacity, lengthening, or expansion of the existing collection, treatment, or dispersal component.

Facultative ponds A wastewater treatment pond that includes surface aeration and algal photosynthesis for oxygen replenishment.

Fecal coliform Bacteria found in the intestinal tracts of warm-blooded animals. The presence of high numbers of fecal coliform bacteria in a water body can indicate the recent release of untreated sewage and/or the presence of animal feces. These organisms may also indicate the presence of pathogens that are harmful to humans.

Fecal typing Assessment technique (e.g., DNA fingerprinting) used to isolate human and non-human sources of fecal coliform contamination in surface and groundwater.

Field capacity The percentage of water remaining in the soil 2 or 3 days after gravity drainage has ceased from saturated conditions.

Filter Device that removes constituents through processes such as sieving, stagnation, adsorption, or absorption; a filter has both area and depth with respect to flow. An apparatus for separating solid or liquid particles from a gas stream in which they are suspended. This apparatus is generally formed of a porous or fibrous layer or of an assembly of porous and/or fibrous layers (by extension, applied also to some oil-bath devices and some electrical devices).

Filter, activated carbon Device filled with a porous form of carbon that is used to decolorize liquids, recover solvents, and remove toxins and odors from water and air.

Filter, bottomless media Media filter that does not incorporate a liner or other physical barrier between the media and the existing soil on which it has been placed; used as a final treatment and dispersal component.

Filter, coir Media filter that uses organic fibric material (coir) from outer husk of coconut.

Filter, disk Device consisting of concentrically grooved disks stacked one upon the other and used for the removal of particles larger than a specific size, typically used in drip distribution systems.

Filter, media Device that uses materials designed to treat effluent by reducing BOD and/or removing suspended solids in an unsaturated environment; biological treatment is facilitated via microbial growth on the surface of the media.

Filter medium The material of which the biological filter is formed and on which a biological film containing bacteria and fungi develops. The part of a filter on or within which the particles are retained.

Filter, peat Media filter that uses appropriate organic fibric material (peat) as the media, typically packaged as prefabricated modular units with the media in a container; a type of biofilter.

Filter, sand Media filter which uses sand of particular specifications as the media.

Filter, screen Filter consisting of a mesh material configured as a cylinder and used to remove particles larger than a specific size in pressurized systems.

Filter strip Strip or area of vegetation often situated at the edge of a field or along a waterway that is used for removing sediment, organic matter, and other pollutants from stormwater runoff.

Filter, trickling Type of media filter which uses a variety of media such as rigid plastics of varying shapes, stone, or tire chips; also includes a clarifier in its configuration and may include a recirculation mode.

Filter, upflow Media filter through which wastewater flows from a lower to a higher elevation, usually characterized by an anaerobic environment.

Filtration The removal of suspended materials using processes such as sieving, stagnation, adsorption, absorption, and possibly biochemical degradation. The separation by a filter of solid particles or liquid particles from a gas stream in which they are suspended (by extension, also the whole of the activities involved in the construction and commissioning of a filter installation).

Filter run The time of filter operation between backwashes.

Final effluent The effluent discharged from a sewage treatment plant.

Final treatment and dispersal Last treatment component (or combination of components) through which effluent is returned to the hydrologic cycle via a soil treatment area or a discharging outfall.

First flush The first portion of runoff, usually defined as a depth in inches, considered to contain the highest pollutant concentration resulting from a rainfall event.

Floc The agglomeration of smaller particles in a gelatinous mass that can be more easily removed from the liquid than the individual small particles.

Flocculant Same as flocculating agent, the catalyst substance that causes the chemical reaction with TSS to form flocculent many times encapsulating the solids.

Flocculent The "floc" or wooly mass of clusters that is formed in flocculation. Many times used interchangeably with "flocculant," however, truly refers to the floc mass and not the catalyst flocculating agent.

Flocculation The provision of retention time with gentle agitation to allow the floc particles or precipitate, associated with the impurities to increase in size by agglomeration.

Flocculating agent The flocculant or chemical used to cause flocculation.

Floodplain For a given flood event, the area of land adjoining a continuous water course that has been covered temporarily by water.

Flow, instantaneous Highest recorded flow occurring within a short, specific period of time (expressed in gallons per minute).

Flow, peak Highest flow occurring within a specified time (minutes, hours, days, etc.); may be further expressed as peak hourly flow, peak daily flow, peak monthly flow, peak seasonal flow, etc.

Flow, surge Flow of effluent that occurs in a short enough period of time that it upsets the function of one or more components of the treatment train.

Flow equalization System configuration that includes sufficient effluent storage capacity to allow for uniform flow to a subsequent component despite variable flow from the source.

Fluid Substances which are transported through a pipeline in liquid and/or gaseous phase.

Fluid conductivity The constant of proportionality in Darcy's law relating the rate of flow of a fluid through a cross-section of porous medium in response to a hydraulic gradient. Fluid conductivity is a function of the intrinsic permeability of a porous medium and the kinematic viscosity of the fluid which flows through it. Fluid conductivity has units of length per time (cm/sec.).

Fly ash Ash entrained by combustion gases. Airborne solid particles that result from the burning of coal and other solid fuels.

Fog A general term applied to a suspension of droplets in a gas. In meteorology, it refers to a suspension of water droplets resulting in the visibility of less than 1 km.

Formaldehyde A chemical compound, the simplest aldehyde, chemical symbol CH_2O. Formaldehyde is a common pollutant, a VOC.

Fossil fuels Fuels such as coal, oil, and natural gas; so-called because they are the remains of ancient plant and animal life.

Free product A contaminant in the unweathered phase, where no dissolution or biodegradation has occurred.

Free water surface wetland (FWS) A lined basin or channel with porous plant substrate and wetland vegetation in which the shallow water is exposed to the air.

Frequency analysis It allows to separate the main components of the signals by dividing the frequency bands using a set of filters.

Frequency of storm Anticipated number of years between storms of equal intensity and/or total rainfall volume. For example, a 25-year 24-hour storm is the volume of rainfall that could be expected to occur during a 24-hour period once every 25 years on average.

Freshwater All waters that would have a chloride ion content of less than 500 parts per million under natural conditions.

Freshwater classifications Class C: freshwaters protected for secondary recreation, fishing, and propagation and survival of aquatic life; all freshwaters are classified to protect these uses at a minimum.Class B: freshwaters protected for primary recreation, which includes swimming on a frequent or organized basis, and all Class C uses.Class WS-I: waters protected as water supplies which are essentially in natural and undeveloped watersheds.Class WS-II: waters protected as water supplies which are generally in predominantly undeveloped watersheds.Class WS-III: waters protected as water supplies which are generally in low to moderately developed watersheds.Class-IV: waters protected as water supplies which are generally in moderately to highly developed watersheds.Class-V: waters protected as water supplies which are generally upstream of and draining to Class-IV waters.

Fugitive dust Dust particles that are introduced into the air through certain activities such as soil cultivation, or vehicles operating on open fields or dirt roadways.

Fugitive emissions Emissions not caught by a capture system that are often due to equipment leaks, evaporative processes, and windblown disturbances.

Fungi Small, non-chlorophyll-bearing plants, without roots, stems, or leaves, which tend to overpower bacteria at low pH and dissolved oxygen concentrations. They generally have a filamentous-type structure and are therefore not welcome in a secondary process clarifier.

Fume An aerosol of solid particles, usually from metallurgical processes, generated by condensation from the gaseous state generally after volatilization from melted substances and often accompanied by chemical reactions such as oxidation.

Fumes In popular usage, gaseous effluents, often unpleasant and malodorous, which might arise from chemical processes.

Gas A mixture of gaseous compounds or elements flowing in a duct, carrying particulate matter.

Gas purifier An apparatus for totally or partially removing one or more constituents from a gas mixture.

Geographic information system (GIS) A computerized database system containing information on natural resources and other factors that can be analyzed and displayed in spatial or map format.

Global warming An increase in the temperature of the Earth's troposphere. Global warming has occurred in the past as a result of natural influences, but the term is most often used to refer to the warming predicted by computer models to occur as a result of increased emissions of greenhouse gases.

Grab sample A sample which is taken from a waste stream on a one-time basis without consideration of the flow rate of the waste stream and without consideration of time.

Grade The slope of a specific surface of interest, such as a road, channel bed or bank, top of embankment, bottom of excavation, or natural ground—commonly measured in percent (unit of measurement per one hundred units) or a ratio of horizontal to vertical distance.

Granular media The material used to effect filtration.

Grassed swale An earthen conveyance system that is broad and shallow with check dams and vegetated with erosion-resistant and flood-tolerant grasses, engineered to remove pollutants from stormwater runoff by filtration through grass and infiltration into the soil.

Gray water Water captured from non-food preparation sinks, showers, baths, spa baths, clothes washing machines, and laundry tubs.

Grease Fats, soaps, oils, waxes, etc., in wastewater.

Greenhouse effect The warming effect of the Earth's atmosphere. Light energy from the sun that passes through the Earth's atmosphere is absorbed by the Earth's surface and reradiated into the atmosphere as heat energy. The heat energy is then trapped by the atmosphere, creating a situation similar to that which occurs in a car with its windows rolled up.

Greenhouse gases Atmospheric gases such as carbon dioxide, methane, chlorofluorocarbons, nitrous oxide, ozone, and water vapor that slow the passage of reradiated heat through the Earth's atmosphere.

Grit Heavy, inorganic matter, such as sand or pebbles. Airborne solid particles in the atmosphere or flues.

Grit chamber Usually in municipal wastewater treatment, a chamber or tank in which primary influent is slowed down so heavy typically inorganic solids can drop out, such as metals and plastics.

Groundwater Subsurface water in the saturation zone from which wells and springs are fed. In a strict sense, the term applies only to water below the water table. Also called "phreatic water" and "plerotic water." Water that fills all of the unblocked pores of material underlying the water table within the upper limit of saturation. Underground water stored in aquifers. Groundwater is created by rain which soaks into the ground and flows down until it collects above an impervious zone. Any water, except capillary moisture, beneath the land surface in the zone of saturation or beneath the bed of any stream, lake, reservoir, or other body of surface water within the boundaries of this commonwealth, whatever may be the subsurface geologic structure in which such water stands, flows, percolates, or otherwise occurs.

Habitat assessment The evaluation of the physical, biological, and chemical environment and evaluation of its impact on biodiversity and ecosystem function and integrity.

Hard liquid A liquid with a vapor pressure below the prevailing atmospheric pressure, e.g., stabilized crude oil.

Harvesting (forestry) All planning and design, road, log deck and skid trail construction, and maintenance during active logging to remove wood products from the forest to a processing plant.

Hazardous air pollutants (HAPs) Toxic chemicals that cause serious health and environmental effects. Health effects include cancer, birth defects, nervous system problems, and death due to massive accidental releases such as occurred at the pesticide plant in Bhopal, India. Hazardous air pollutants are released by sources such as chemical plants, dry cleaners, printing plants, and motor vehicles (cars, trucks, buses, etc.).

Hazardous waste Hazardous waste is solid waste which poses specified health and environmental hazards.

Headworks The beginning of the treatment plant where the influent begins treatment.

Heavy metals Metals which can be precipitated by hydrogen sulfide in an acid solution, including lead, silver, gold, mercury, bismuth, and copper. Those metals that have high specific gravity and high atomic mass, such as lead, cadmium, zinc, copper, silver, and mercury. In sufficient concentrations, these metals are toxic to humans and aquatic life.

Henry's law The relationship between the partial pressure of a compound and the equilibrium concentration in the liquid through a constant of proportionality known as Henry's law constant. See partial pressure.

Heterogeneous Varying in structure or composition and having different properties in different locations or directions.

High-priority watershed A watershed assigned to the category of nominal scaling associated with the greatest impacts to water quality for the criteria being ranked (i.e., total agriculture high-priority watershed). When the criteria are not explicitly referenced with this term, the highest rank of the overall (total) NPS pollution assignment is implied.

Hood An inlet device for an extraction system.

Hydrocarbons Compounds containing various combinations of hydrogen and carbon atoms. They may be emitted into the air by natural sources (e.g., trees) and as a result of fossil and vegetative fuel combustion, fuel volatilization, and solvent use. Hydrocarbons are a major contributor to smog.

Hydrochlorofluorocarbon (HCFC) A chemical compound that would be a hydrocarbon except that one or more hydrogen atoms in each molecule are replaced by a chlorine atom and one or more hydrogen atoms are replaced by a fluorine atom. Some HCFCs are implicated in the destruction of stratospheric ozone.

Hydrofluorocarbon (HCA) A chemical compound that would be a hydrocarbon except that one or more hydrogen atoms in each molecule are replaced by a fluorine atom.

Hydrogen ion concentration [H+] The weight of hydrogen ion in moles per liter of solution. Commonly expressed as the pH value, which is the logarithm of the reciprocal of the hydrogen ion concentration.

Hydrogen sulfide gas (H_2S) Hydrogen sulfide is a gas with a rotten egg odor. This gas is produced under anaerobic conditions. Hydrogen sulfide gas is particularly dangerous because it dulls the sense of smell so that you do not notice it after you have been around it for a while. In high concentrations, hydrogen sulfide gas is only noticeable for a very short time before it dulls the sense of smell. The gas is very poisonous to the respiratory system, explosive, flammable, colorless, and heavier than air.

Hydraulic conductivity The constant of proportionality in Darcy's law relating the rate of flow of water through a cross-section of porous medium in response to a hydraulic gradient. Also known as the coefficient of permeability, hydraulic conductivity is a function of the intrinsic permeability of porous medium and the kinematic viscosity of the water which flows through it. Hydraulic conductivity has units of length per time (cm/sec.).

Hydraulic gradient The change in piezometric head between two points divided by the horizontal distance between the two points, having dimensions of length per length (cm/cm). See piezometric head.

Hydrologic cycle The movement of water in and on the earth and atmosphere through processes such as precipitation, infiltration, runoff, and evaporation.

Hydrolysis Conversion of organic nitrogen to ammonia by enzymes secreted by bacteria, plants, and animals in a reaction that adds water.

I&A Innovative and alternative. A term defined by the EPA to describe non-conventional technologies. "Alternative systems" are fully proven systems that reclaim or reuse wastewater, productively recycle wastewater components, recover energy, or eliminate the discharge of pollutants. A variety of wastewater treatment systems have been included in the definition, including land treatment, aquaculture, containment ponds, and on-site treatment using small diameter or vacuum sewers.

IAQ Indoor air quality

Immediate oxygen demand (IOD) IOD is the amount of oxygen that is utilized by the components of a wastewater within 15 minutes (unless otherwise specified) after being introduced into water that contains dissolved oxygen.

Impact A collision of two particles with each other or of a particle with a solid or liquid surface.

Impaction The action of particles entering into contact with a surface.

Impaired water Water that is not meeting the state water quality standard; water with fish or shellfish harvesting prohibition by the Virginia Department of

Health (VDH); water where biological monitoring indicates moderate or severe impairment.

Impermeable Not permitting the passage of fluid through pores; in practical terms, some small level of hydraulic conductivity may occur, but at so low a level (e.g., 1×10^{-7} cm/s) that it is considered to be negligible.

Impervious cover A surface composed of any material that significantly impedes or prevents natural infiltration of water into soil—includes (but not limited to) roofs, buildings, streets, parking areas and any concrete, asphalt, or compacted gravel surface.

Impoundment An artificial collection or storage of water, as a reservoir, pit, dugout, sump, etc.

Indirect source control program Rules, regulations, local ordinances and land use controls, and other regulatory strategies of air pollution control districts or local governments used to control or reduce emissions associated with new and existing indirect sources. Indirect source control programs include regulatory strategies such as transportation control measures; parking charges; land use controls that reduce the need for vehicle travel and increase transit, bicycle, and pedestrian access; and source-specific regulations such as truck idling and travel schedule requirements.

Indoor air pollution Air pollutants that occur within buildings or other enclosed spaces, as opposed to those occurring in outdoor, or ambient air. Some examples of indoor air pollutants are nitrogen oxides, smoke, asbestos, formaldehyde, and carbon monoxide.

Industrial ecology IE focuses on combining perpetually desirable outcomes in environment, economy, and technology sustainably. The primary tenet is that all systems mimic nature and are thus closed loop, continuous, circular. In wastewater treatment, industrial ecology would mean that all so-called waste is re-input into the same or other processes. For example, biosolids as fertilizer can be considered a use of sludge consistent with industrial ecology. Recycling wastewater into the treatment plant, manufacturing, or the other process is another example.

Industrial sources Non-municipal or industrial sources often generate wastewater that is discharged to surface waters. The types of wastewaters generated at a facility depend on the specific activities undertaken at a particular site and may include manufacturing or process wastewaters, cooling waters, sanitary wastewater, and stormwater runoff.

Industrial wastes The solid and liquid wastes originating from industrial processes

Industrial wastewater treatment Wastewater treatment for industries such as manufacturing, food processing, corrugators, printing. Paper and pulp mills'

treatment of wastewater is an example of industrial wastewater treatment. Municipal wastewater treatment would be an example not considered to be industrial.

Infiltration The seepage of groundwater into a sewer system, including service connections. Seepage frequently occurs through defective or cracked pipes, pipe joints and connections, interceptor access risers and covers, or manhole walls. The downward movement of water through a soil from rainfall or from the application of artificial recharge in response to gravity and capillarity.

Infiltration facility A stormwater management facility that temporarily impounds runoff and discharges it via infiltration through the surrounding soil—may be equipped with an outlet structure to discharge impounded runoff, such discharge is normally reserved for overflow and other emergency conditions (infiltration basin, infiltration trench, infiltration dry well, and porous pavement are considered infiltration facilities).

Infiltration/Inflow The total quantity of water from both infiltration and inflow without distinguishing the source.

Inflow Water discharged into a sewer system and service connections from such sources as, but not limited to, roof leaders, cellars, yard and area drains, foundation drains, cooling water discharges, drains from springs and swampy areas, around manhole covers or through holes in the covers, cross-connections from storm and combined sewer systems, catch basins, stormwaters, surface runoff, street wash waters, or drainage. Inflow differs from infiltration in that it is a direct discharge into the sewer rather than a leak in the sewer itself.

Influent Water, wastewater, or other liquids—raw (untreated) or partially treated—flowing into an interceptor, reservoir, basin, treatment process, or treatment plant. The untreated wastewater or raw sewage coming into a wastewater treatment plant.

Influent screens Screens used to remove large inorganic solids from the waste stream.

Infrasonics The sound of frequency less than 20 HZ.

Inhalable particles All dust capable of entering the human respiratory tract.

Inhibitory substances Materials that kill or restrict the ability of organisms to treat wastes.

Injection well Well by which effluent is transmitted to an underground formation; in most cases, these are regulated and require a permit from a regulatory authority.

Inorganic Material such as sand, salt, iron, calcium salts, and other mineral materials. Inorganic substances are of mineral origin, whereas organic substances are usually of animal or plant origin. Also see "organic."

Inorganic waste Waste material such as sand, salt, iron, calcium, and other mineral materials which are only slightly affected by the action of organisms. Inorganic wastes are chemical substances of mineral origin, whereas organic wastes are chemical substances of an animal or plant origin.

Inorganic material Material that will not respond to biological action (sand, cinders, stone). Non-volatile fraction of solids.

In situ treatment In situ treatment means treatment of the soil in place, i.e., they are not dug up. The excavation of contaminated soils usually adds significant expense to a site remediation.

Intensity Depth of rainfall divided by duration.

Interfacial tension Phenomena occurring at the interface of a liquid and gas where the liquid behaves as it if were covered by an elastic membrane in a constant state of tension. The tension is due to unbalanced attractive forces between the liquid molecules at the liquid surface.

Internal combustion engine An engine in which both the heat energy and the ensuing mechanical energy are produced inside the engine. This includes gas turbines, spark ignition gas, and compression ignition diesel engines.

Inversion A layer of warm air in the atmosphere that prevents the rise of cooling air and traps pollutants beneath it.

Irritant A substance that causes irritation of the skin, eyes, or respiratory system. Effects may be acute from a single high-level exposure, or chronic from repeated low-level exposures to such compounds as chlorine, nitrogen dioxide, and nitric acid.

Isokinetic sampling Sampling at such a rate that the average velocity of the gas entering the sample nozzle is the same as that of the gas in the duct at the sampling point.

JTU The Jackson Turbidity Unit (JTU) is a measurement of the cloudiness in water caused by stirred sediments.

Karst topography Regions that are characterized by formations underlain by carbonate rock typified by the presence of limestone caverns and sinkholes.

L10, L50 and L90 The sound levels exceeding 10, 50, and 90 % of the total time intervals during a particular period are designated as L10, L50, and L90, respectively.

Lagoon Constructed basin lined with either soils with very low permeability or a synthetic material, surrounded with berms and which contains at least three feet of wastewater which utilizes sunlight, wind or mechanical aeration, and natural bacteria to break down waste via physical, chemical, and biological processes.

Lagoon, evaporation Lagoon where wastewater is stored and the water is allowed to evaporate over time.

Lagoon, storage Lagoon where some form of wastewater is stored before it is either conveyed to another component for further processing or is reused.

Land conversion Final harvest of the forest with subsequent land use conversion to agriculture, residential or commercial development, mining or highway construction.

Land development characteristics A man-made change to, or construction on, the land surface that changes its runoff characteristics.

Land use Any activities that takes place on land, such as construction, farming, or tree removal.

Laterals Laterals are sewers collecting the effluent from two or more sublaterals discharging to "Mains."

Lagoon sludge Lagoon sludge is a relatively shallow basin, or natural depression, used for the storage or digestion of sludge, sometimes for its ultimate detention or dewatering.

LC50 The concentration of a toxicant or percentage dilution of an effluent that is predicted to be lethal to 50 % of a test population of organisms.

Ldn The day–night equivalent value of sound level. The day is counted from 6 a.m. to 9 p.m. (15 h) and night from 9 p.m. to 6 a.m. (9 h).

Leak An uncontrolled fluid release from a pipeline and other sources.

Leak consequences The result of a pipeline leak and other sources in terms of human safety and damage to the environment. Economic loss such as cost of repair and deferred production are not taken into account in the leak consequence evaluation methodology given in this book.

Leak expectancy The probability of occurrence of a leak.

Lime Any of several compounds consisting of calcium hydroxide ($Ca(OH)_2$) or calcium oxide (CaO).

Liquefaction Liquefaction as applied to sludge digestion means the transformation of large solid particles of sludge into either a soluble or a finely dispersed state.

Liquid solids separation The process of separating the liquids and solids in a given wastewater. Liquid/solids separation comes in one of the 3 processes: 1. If the solids sink (specific gravity greater than 1), use a clarifier; 2. if the solids float (specific gravity less than 1), use a floatation unit (DAF); 3. if neither sink or float (specific gravity is 1), try using a screen (rotary or parabolic).

Load or loading The introduction of an amount of matter or thermal energy into a receiving water; may be either caused by man (pollutant loading) or natural (background loading).

Loading Quantity of material applied to a device at one time. Amount of a substance entering the environment (soil, water, or air).

Loading rate, areal Quantity of effluent applied to the footprint of the soil treatment area (or the absorption area of an above-grade soil treatment area) expressed as volume per area per unit time, e.g., gallons per day per square foot (gpd/sq. ft.).

Loading rate, biochemical Quantity of BOD5 delivered to a treatment component expressed as mass per time (e.g., pounds of BOD5 per day).

Loading rate, biological Quantity of organic matter delivered to a treatment component expressed mass per time (e.g., pounds per day).

Loading rate, mass Sum of organic and inorganic effluent constituents delivered to a treatment component in a time interval, expressed as mass per time.

Loading rate, nutrient Sum of organic and inorganic nutrients (primarily nitrogen and phosphorus) delivered to a treatment component in a specified time interval expressed as mass per time.

Loading rate, organic Biodegradable fraction of chemical oxygen demand (biochemical oxygen demand, biodegradable FOG, and volatile solids) delivered to a treatment component in a specified time interval expressed as mass per time or area; e.g., pounds per day or pounds per cubic foot per day (pretreatment); pounds per square foot per day (infiltrative surface or pretreatment); typical residential system designs assume biochemical loading equals organic loading.

Local limits Conditional discharge limits imposed by municipalities upon industrial or commercial facilities that discharge to the municipal sewage treatment system.

Lowest achievable emission rate (LAER) Considered to be the lowest rate of emissions from a source category which is contained in the State Implementation Plan or which is achieved in practice by such category of sources. This term is most often associated with a non-attainment area.

Main sewer A sewer line that receives wastewater from many tributary branches and sewer lines and serves as an outlet for a large territory or is used to feed an intercepting sewer.

Maintenance (forestry) Maintenance includes upkeep of permanent road and trail systems, prescribed burning for fuel reduction or habitat selection, and use of herbicides.

Marsh A wet area, periodically inundated with standing or slow-moving water, that has grassy or herbaceous vegetation and often little peat accumulation; the water may be salt, brackish, or fresh.

Makeup water Water supplied to replenish the water of a system.

Malfunction Condition in which a component is not performing as designed/installed.

Manholes Manholes are used in sewer mains as junction points and sediment traps and to provide access for maintenance and inspection.

Material safety data sheets (MSDSs) Product safety information sheets prepared by manufacturers and marketers of products containing toxic chemicals. These sheets can be obtained by requesting them from the manufacturer or marketer. Some stores, such as hardware stores, may have material safety data sheets on hand for products they sell.

Maximum achievable control technology (MACT) Federal emission limitations based on the best demonstrated control technology or practices in similar sources to be applied to major sources emitting one or more federal hazardous air pollutants.

Media The material in a trickling filter on which slime accumulates and organisms grow. As settled wastewater trickles over the media, organisms in the slime remove certain types of wastes, thereby partially treating the wastewater. Also the material in a rotating biological contactor or in a gravity or pressure filter.

micro- (Âµ) The metric prefix for one millionth of the unit that follows.

microgram (Âµg) One millionth of a gram: 1 Âµg = 10^{-6} g = 0.001 mg.

micrograms per cubic meter (Âµg/m^3) The mass in micrograms of a substance contained within a cubic meter of another substance or vacuum. This is the standard unit of measure for the mass density (concentration) of particles suspended in air; also sometimes used for the concentration of gases in air.

micrometer (Âµm), micron One millionth of a meter: 1 Âµm = 10^{-6} m.

Microorganisms Very small organisms that can be seen only through a microscope. Some microorganisms use the wastes in wastewater for food and thus remove or alter much of the undesirable matter.

MIST A suspension of droplets in a gas.

Mixed liquor The combination of primary effluent and active biological solids (return sludge) in the activated sludge process that is fed into the aeration tank.

Mixed liquor suspended solids (MLSS) Suspended solids in the mixed liquor of an aeration tank.

Mixed liquor volatile suspended solids (MLVSS) The organic or volatile suspended solids in the mixed liquor of an aeration tank. The volatile portion is used as a measure or indication of the microorganisms present.

Mixing zone An area where an effluent discharge undergoes initial dilution and is extended to cover the secondary mixing in the ambient water body. A mixing zone is an allocated impact zone where water quality criteria can be exceeded as long as acutely toxic conditions are prevented.

Mixed media gravity filter A filter using more than one filtering media (such as coal and sand).

Moisture content The amount of water or lost from the soil upon drying to a constant weight, expressed as the weight per unit weight of dry soil or as the volume of water per unit bulk volume of the soil. For a fully saturated medium, moisture content equals the porosity; in the vadose zone, moisture content ranges between zero and the porosity value for the medium. See porosity, vadose zone, saturated zone.

Molecular diffusion Process where molecules of various gases tend to intermingle and eventually become evenly dispersed.

Molecular weight The amount of mass in mole of molecules of a substance determined by summing the masses of the individual atoms comprising the molecule. One mole is equivalent to $6.02 \times 1{,}023$ the molecules.

Monitoring The physical, chemical, and biological analysis of water quality parameters as well as predictive measures of assessing non-point source water quality impacts.

Multimedia exposure Exposure to a toxic substance from multiple pathways such as air, water, soil, food, and breast milk.

Municipal discharge Discharge of effluent from wastewater treatment plants operated by municipalities or public sewerage authorities; may include wastewater from households, commercial establishments, and industries.

Natural systems Ecologically based biological wastewater treatment systems such as constructed wetlands having minimal dependence on mechanical elements.

N: Nitrogen This nutrient is present in various forms in wastewater, principally ammonia and nitrate; an element essential to the growth and development of plants; occurs in manure and chemical fertilizer and, in excess, can cause waters to become polluted by promoting excessive growth of algae and other aquatic plants.

Neutralization Addition of an acid or alkali (base) to a liquid to cause the pH of the liquid to move toward a neutral pH of 7.0.

New source review (NSR) New source review (NSR) typically means any new source locating in an, e.g., ozone non-attainment area that will emit volatile organic compounds (VOCs) and/or oxide of nitrogen (NO) in certain amounts. These sources must undergo a new source review that provides for offsetting emissions for any increases in the emissions of these two pollutants; use the lowest achievable emission technology to control emissions; apply for a construction permit; and meet other state requirements before the new emission from the source can be permitted. Existing sources, located in the ozone non-attainment area, that emit these two pollutants and plan to change their operational methods that will cause an increase in the emissions of these two pollutants must apply for a modification permit and under go a review similar to a new source.

Nitrification An aerobic process in which bacteria change the ammonia and organic nitrogen in wastewater into oxidized nitrogen (usually nitrate). The second-stage BOD is sometimes referred to as the "nitrification stage" (first-stage BOD is called the "carbonaceous stage").

Nitrification stage A stage of decomposition that occurs in biological treatment processes when aerobic bacteria, using dissolved oxygen, change nitrogen compounds (ammonia and organic nitrogen) into oxidized nitrogen (usually nitrate). The second-stage BOD is sometimes referred to as the "nitrification stage" (first-stage BOD is called the "carbonaceous stage").

Nitrifying bacteria Bacteria that change the ammonia and organic nitrogen in wastewater into oxidized nitrogen (usually nitrate).

Nitrobacteria Principal genera of autotrophic bacteria responsible for the second step of biological nitrification: conversion (oxidation) of nitrite to nitrate.

Nitrogen fixation The conversion of nitrogen gas to organic nitrogen, ammonia, or nitrate. Nitrogen fixation can occur biologically (i.e., conversion of nitrogen gas to organic nitrogen by certain photosynthetic blue-green algae), by natural physical processes (i.e., conversion of nitrogen gas to nitrate by lightning), or by industrial processes (manufacture of fertilizers and explosives).

Nitrogenous A term used to describe chemical compounds (usually organic) containing nitrogen in combined forms. Proteins and nitrates are nitrogenous compounds.

Nitrogen oxides (NOx) A criteria air pollutant. Nitrogen oxides are produced from burning fuels, including gasoline and coal. Nitrogen oxides are smog-formers, which react with volatile organic compounds to form smog. Nitrogen oxides are also major components of acid rain.

Noise Noise is unwanted sound.

Non-aqueous phase liquid (NAPL) Contaminants that remain as the original bulk liquid in the subsurface.

Non-attainment area A geographic area in which a criteria air pollutant level is higher than allowed by the federal standards. A single geographic area may have an acceptable level for one criteria air pollutant, but have unacceptable levels of one or more other criteria air pollutants. Thus, an area can be both an attainment and non-attainment area at the same time. Sixty percent of Americans are estimated to live in non-attainment areas.

Non-point sources Diffuse pollution sources that are not recognized to have a single point of origin; diffused pollutants that are washed off the land (runoff) during the natural process of rainwater flowing across the land to rivers, lakes, oceans, and other water bodies.

Non-point source assessment An evaluation of the state's waters on a watershed basis, consisting of the calculation of ordinal values for a number of NPS-pollution-related water quality impacting criteria and resulting in (1) the nominal scaling of these criteria measures into three ranks and (2) the creation of an overall NPS pollution water quality assignment similarly ranked.

Non-point source pollution Sources of water pollution not associated with a distinct discharge source; includes rainwater erosion, runoff from roads, farms, parking lots, and seepage from soil-based wastewater disposal systems.

Nutrient Any substance that is assimilated (taken in) by organisms and promotes growth. Nitrogen and phosphorus are nutrients which promote the growth of algae. There are other essential and trace elements which are also considered nutrients.

Nutrient cycle The transformation or change of a nutrient from one form to another until the nutrient has returned to the original form, thus completing the cycle.

Nutrient sensitive waters Waters subject to excessive growth of microscopic and macroscopic vegetation that needs additional nutrient management. In general, management strategies for point and non-point source pollution control are designed to prevent any increase in nutrients over background levels.

Octane number A numerical measure of the antiknock properties of gasoline used as a motor fuel. The higher the octane number, the greater the antiknock properties.

Octanol-water partition coefficient (Kow) A coefficient representing the ratio of solubility of a compound in octanol to its solubility in water. As Kow increases, water solubility decreases.

Odor Quality of gases, liquids, or particulates that stimulates the olfactory organ.

Oil interceptor Oil interceptor is a device designed to remove small oil globules by gravity from the water by limiting the flow velocity and the overflow rate.

Onsite Local wastewater treatment for a single house or small community.

Opacity The amount of light obscured by particle pollution in the atmosphere. Opacity is used as an indicator of changes in performance of particulate control systems.

Organic Substances that come from animal or plant sources. Organic substances always contain carbon (inorganic materials are chemical substances of mineral origin).

Organic compounds A large group of chemical compounds containing mainly carbon, hydrogen, nitrogen, and oxygen. All living organisms are made up of organic compounds.

Organic loading The pounds of BOD per day applied to a unit process.

Organic waste Waste material which comes mainly from animal or plant sources. Organic wastes generally can be consumed by bacteria and other microscopic organisms. Inorganic wastes are chemical substances of mineral origin.

Organics 1. A term used to refer to chemical compounds made from carbon molecules. These compounds may be natural materials (such as animal or plant sources) or man-made materials (such as synthetic organics). Also see "organic." 2. Any form of animal or plant life.

Outfall Place where effluent is discharged into receiving waters.

Outstanding resource waters unique waters of exceptional state or national recreational or ecological significance that requires special protection to maintain existing uses.

Overland flow land treatment Partially treated wastewater is applied to relatively impermeable soils at the top of a grass-covered gradient. The waste is cleaned by the vegetation and microbial action, and excess water is captured at the bottom of the slope.

Oxidation The conversion of organic material to a more stable form using bacteria, chemicals, or oxygen.

Oxidation ponds Oxidation ponds are basins in which wastewater undergoes a biological oxidation treatment by action of algae and bacteria.

Oxidation direct Oxidation direct is oxidation of substances in sewage without the benefit of living organisms, by the direct application of air or oxidizing agents such as chlorine.

Oxidation sewage Oxidation sewage is the process whereby, through the agency of living organisms in the presence of oxygen, the organic matter that is contained in sewage is converted into a more stable or a mineral form.

Oxygen consumed Oxygen consumed is the quantity of oxygen taken up from potassium permanganate in solution by a liquid containing organic matter.

Commonly regarded as an index of the carbonaceous matter present. Time and temperature must be specified.

Oxygen demand Chemical oxygen demand and biological oxygen demand (COD and BOD) are measures of the oxygen consumed when a substance degrades. Materials such as food waste and dead plant or animal tissue use up dissolved oxygen in the water when decomposed through chemical or biological processes.

Ozone A gas which is a variety of oxygen. The oxygen gas found in the air consists of two oxygen atoms stuck together; this is molecular oxygen. Ozone consists of three oxygen atoms stuck together into an ozone molecule. Ozone occurs in nature; it produces the sharp smell you notice near a lightning strike. High concentrations of ozone gas are found in a layer of the atmosphere—the stratosphere—high above the Earth. Stratospheric ozone shields the Earth against harmful rays from the sun, particularly ultraviolet B. Smog's main component is ozone; this ground-level ozone is a product of reactions among chemicals produced by burning coal, gasoline and other fuels, and chemicals found in products such as solvents, paints, and hair sprays.

Ozone depletion The reduction in the stratospheric ozone layer. Stratospheric ozone shields the Earth from ultraviolet radiation. The breakdown of certain chlorine- and/or bromine-containing compounds that catalytically destroy ozone molecules in the stratosphere can cause a reduction in the ozone layer.

P (Phosphorus) This nutrient, which is present in wastewater, acts as a fertilizer for algae in surface waters.

Particle A small discrete mass of solid or liquid matter.

Particle size analysis The science which deals with the measurement of the dimensions and determination of the shape of particles.

Particle size analysis; GRANULOMETRIC ANALYSIS The whole of the operations by which a particle size (granulometric) distribution may be obtained.

Particle size distribution; GRANULOMETRIC DISTRIBUTION A presentation, in the form of tables of numbers or of graphs, of the experimental results obtained using a method or an apparatus capable of measuring the equivalent diameter of particles in a sample or capable of giving the proportion of particles for which the equivalent diameter lies between defined limits.

Particulates Solid matter, in a gas stream, that is solid at normal temperature and pressure.

Particulates: particulate matter (PM-10) Particulate matter is a criteria air pollutant and is a finely divided particle with an aerodynamic diameter of 10 micrometers or less. Particulate matter includes dust, soot, and other tiny bits of solid materials that are released into and move around in the air. Particulates are produced by many sources, including burning of diesel fuels by trucks and buses, incineration of garbage, mixing and application of fertilizers and pesticides, road construction, industrial processes such as steel making, mining operations, agricultural burning (field and slash burning), and operation of fireplaces and wood stoves. Particulate pollution can cause eye, nose, and throat irritation and other health problems.

Particulates: particulate matter (PM2.5) Includes tiny particles with an aerodynamic diameter less than or equal to a nominal 2.5 microns. This fraction of particulate matter penetrates most deeply into the lungs.

Parts per billion (ppb)/parts per million (ppm) Units commonly used to express contamination ratios, as in establishing the maximum permissible amount of contaminant in water, land, or air.

Parts per million (ppm) Parts per million (ppm) is parts by mass in sewage analysis; ppm by mass is equal to milligrams per liter divided by the relative density (specific gravity). In water analysis, ppm is always understood to imply mass/mass ratio (mg/kg), even though in practice a volume may be measured instead of a mass.

Pathogenic Disease causing or harmful to man.

Pathogenic organisms Organisms, including bacteria, viruses or cysts, capable of causing diseases (giardiasis, cryptosporidiosis, typhoid, cholera, dysentery) in a host (such as a person). There are many types of organisms which do not cause disease. These organisms are called non-pathogenic.

Pathogen Disease-causing biological agent such as a bacterium, virus, or fungus.

Peak demand The maximum momentary load placed on a water treatment plant, pumping station, or distribution system. This demand is usually the maximum average load in one hour or less, but may be specified as the instantaneous load or the load during some other short time period.

Peak levels A level of airborne pollutants that is much higher than average. They can occur over a short period of minutes or hours in response to sudden releases, or they can occur due to a longer-term build-up over several days.

Peaking factor Ratio of a maximum flow to the average flow, such as maximum hourly flow or maximum daily flow to the average daily flow.

Penetration The ratio of the quantity of particles leaving a filter, dust separator, or a droplet separator, to the quantity entering it.

Percent saturation The amount of a substance that is dissolved in a solution compared with the amount that could be dissolved in solution, expressed as a percent.

Perceived noise levels (PNL) The noise of aircraft is described in terms of perceived noise levels (PNL), a scale of noisiness, expressed in pNdB.

Percolation The flow of liquid through a filtering medium.

Performance standard A limitation on the emission or discharge of a pollutant that may be expressed as an emission or discharge standard or as a requirement for specific operating procedures.

Permeability Ability of a porous medium such as soil to transmit fluids (liquids or gases). A measure of a soil resistance to fluid flow. Permeability, along with fluid viscosity and density, is used to determine fluid conductivity.

pH A measure of acidity or alkalinity of water, or any given substance. The scale is 1–14 with 7 being neutral. Over 7 is alkaline or caustic and under 7 is acid or base; numerical measure of hydrogen ion activity with a scale of 0–14. Neutral is pH 7; values below 7 are acid and values above 7 are alkaline. Waters that are too acid or alkaline can be unfit for animal or plant life.

Phase The physical form in which a substance is found. The three major phases are liquid, vapor, and dissolved in pore water.

Phosphorus (P) Chemical element and nutrient essential for all life forms, occurring as orthophosphate, pyrophosphate (P_2O_7-4), tripolyphosphate ($P_3O_{10}5-$), and organic phosphate forms; each of these forms, as well as their sum (total phosphorus), is expressed in terms of milligrams per liter (mg/L) elemental phosphorus and occurs in natural waters and wastewater almost solely as phosphates; excess levels of phosphorous in fresh surface waters may contribute to eutrophication.

Phosphorus, inorganic Forms of phosphorus from mineral sources, such as orthophosphate, pyrophosphate (P_2O_7-4), and tripolyphosphate ($P_3O_{10}5-$).

Phosphorus, organic Phosphorus formed primarily by biological processes; sources of organic phosphorus in sewage include bodily wastes, food residues, and the conversion of orthophosphates in biological treatment processes.

Physical waste treatment process Physical waste treatment processes include use of racks, screens, comminutors, clarifiers (sedimentation and flotation), and filtration. Chemical or biological reactions are important treatment processes, but not part of a physical treatment process.

Physico-chemical treatment A combination of physical and chemical treatment to achieve a specific result.

Phosphorous (P) An element essential to the growth and development of plants; occurs in manure and chemical fertilizer and, in excess, can cause waters to become polluted by promoting excessive growth of algae and other aquatic plants.

Piezometric head The level to which water from a given aquifer will rise by hydrostatic pressure. For the uppermost unconfined aquifer, the piezometric head is identical to the water table elevation. For confined aquifers, the piezometric head can be above or below the water table.

Pipeline A system of pipes and other components used for the transportation of fluids, between (but excluding) plants. A pipeline extends from pig trap to pig trap (including the pig traps and associated pipework and valves), or, if no pig trap is fitted, to the first isolation valve within the plant boundaries or a more inward valve if so nominated.

Plasticulture Agricultural production practice, usually used with fruits and vegetables, that cover a large portion of the field with plastic to heat up the soil and control weeds; land is graded so that stormwater is quickly drained to minimize excessive soil moisture levels.

Plume A visible or measurable discharge of a contaminant from a given point of origin that can be measured according to the Ringelmann scale.

Point source pollution A specific discharge that is traceable to a distinct source (pipe, ditch, container, well, etc.) such as those from wastewater treatment plants or industrial facilities. Discharges of treated or untreated effluent from industries, wastewater treatment plants, and other sources that can be traced back to a single point of discharge. Some sources (leaching landfills, hazardous wastes, brownfields, material storage, airport deicing, underground storage tanks, etc.) are subject to question, as to whether they fall into the point or non-point source category. In these situations, where NPDES permitting applies, the state of Virginia considers the issue a point source pollution problem, and the topic is not addressed in this non-point source pollution management plan.

Pollutant Any substance which causes impairment (reduction) of water quality to a degree that has an adverse effect on any beneficial use of the water.

Pollutant, conservative Pollutants that do not readily degrade in the environment and which are mitigated primarily by natural stream dilution after entering receiving water bodies. Included are pollutants such as metals.

Pollutant, Non-conservative Pollutants that are mitigated by natural biodegradation or other environmental decay or removal processes in the receiving stream after in-stream mixing and dilution have occurred.

Pollutants (pollution) Unwanted chemicals or other materials found in the air. Pollutants can harm health, the environment, and property. Many air pollutants occur as gases or vapors, but some are very tiny solid particles: dust, smoke, or soot.

Pollution The impairment (reduction) of water quality by agricultural, domestic, or industrial wastes (including thermal and radioactive wastes) to a degree that the natural water quality is changed to hinder any beneficial use of the water or render it offensive to the senses of sight, taste, or smell or when sufficient amounts of wastes create or pose a potential threat to human health or the environment.

Population equivalent The equivalent, in terms of a fixed population, of a varying or transient population, of a hospital or restaurant, based upon a figure of 0.06 kg BOD per head per 120 L per head per day.

Porosity 1. Open space or interstices in rock, other earth materials, or synthetic media; 2. ratio of the open space to the total volume often described as a percentage; 3. the volume fraction of a rock or unconsolidated sediment not occupied by solid material but usually occupied by water and/or air. Porosity is a dimensionless quantity.

Porous layer A permeable layer of solid material in any form having interstices of small size, generally known as "pores."

Postdenitrification Biological wastewater treatment process for nitrogen removal that utilizes an anoxic zone located at the effluent end of an aeration tank. Due to lack of organic carbon, methanol addition is typically required.

Postchlorination The addition of chlorine to the plant effluent, following plant treatment, for disinfection purposes.

Postharvest site restoration All road, deck, and skid trail restoration activities, mechanical site preparation, prescribed burning to remove logging debris, and tree planting to facilitate reforestation of the logged site.

Potable water Water that does not contain objectionable pollution, contamination, minerals, or infective agents and is considered satisfactory for drinking.

Preaeration The addition of air at the initial stages of treatment to freshen the wastewater, remove gases, add oxygen, promote flotation of grease, and aid coagulation.

Prechlorination (wastewater) The addition of chlorine in the collection system serving the plant or at the headworks of the plant prior to other treatment processes mainly for odor and corrosion control. Also applied to aid disinfection, to reduce plant BOD load, to aid in settling, to control foaming in Imhoff units, and to help remove oil.

Precipitation When a substance dissolved in a liquid passes out of solution and into solid form. An operation in which particles are separated from a gas stream in which they are suspended, by the action of an electrical field or a thermal gradient.

Precursor, THM Natural organic compounds found in all surface and groundwater. These compounds may react with halogens (such as chlorine) to form trihalomethanes (THMs); they must be present in order for THMs to form.

Predenitrification Biological wastewater treatment process for nitrogen removal that utilizes an anoxic zone located at the influent end of an aeration tank. Organic matter present in the wastewater serves as a carbon source for denitrifying bacteria.

Preliminary treatment The removal of metal, rocks, rags, sand, eggshells, and similar materials which may hinder the operation of a wastewater treatment plant. Preliminary treatment is accomplished by using equipment such as racks, bar screens, comminutors, and grit removal systems.

Pressure gradient A pressure differential in a given medium, which tends to induce movement from areas of higher pressure to areas of lower pressure.

Pretreatment Any component or combination of components that provides treatment of wastewater prior to conveyance to a final treatment and dispersal component or reuse; often, this treatment is designed to meet primary, secondary, tertiary, and/or disinfection treatment standards.

Pretreatment facility Industrial wastewater treatment plant consisting of one or more treatment devices designed to remove sufficient pollutants from wastewaters to allow an industry to comply with effluent limits established by the US EPA General and Categorical Pretreatment Regulations or locally derived prohibited discharge requirements and local effluent limits. Compliance with effluent limits allows for a legal discharge to a POTW.

Preventive maintenance Regularly scheduled servicing of machinery or other equipment using appropriate tools, tests, and lubricants. This type of maintenance can prolong the useful life of equipment and machinery and increase its efficiency by detecting and correcting problems before they cause a breakdown of the equipment.

Primary clarifier A wastewater treatment device which consists of a rectangular or circular tank that allows those substances in wastewater that readily settle or float to be separated from the wastewater being treated.

Primary settlement tank A tank in which the majority of settable solids are removed from the crude sewage flowing through it.

Primary standard A pollution limit based on health effects. Primary standards are set for criteria air pollutants.

Primary treatment Primary treatment is water purification based on the difference in density of the polluting substance and the medium, the former being removed either by rising or by settling. This process can include screening, grit removal, sedimentation, sludge dig. A wastewater treatment process that takes place in a rectangular or circular tank and allows those substances in wastewater that readily settle or float to be separated from the water being treated.

Primary wastewater treatment The removal of sand, grit, and larger solids from wastewater by screens, settling tanks and/or skimming devices.estion, and sludge disposal.

Priority pollutants The EPA has proposed a list of 126 priority toxic pollutants. These substances are an environmental hazard and may be present in water. Because of the known or suspected hazards of these pollutants, industrial users of the substances are subject to regulation. The toxicity to humans may be substantiated by human epidemiological studies or based on effects on laboratory animals related to carcinogenicity, mutagenicity, teratogenicity, or reproduction. Toxicity to fish and wildlife may be related to either acute or chronic effects on the organisms themselves or to humans by bioaccumulation in food fish. Persistence (including mobility and degradability) and treatability are also important factors.

Process variable A physical or chemical quantity that is usually measured and controlled in the operation of a wastewater treatment plant or an industrial plant.

Propeller fan Fan in which the air enters and leaves the impeller in a direction substantially parallel to its axis.

Protozoa A group of motile microscopic animals (usually single-celled and aerobic) that sometime cluster in colonies and often consume bacteria as an energy source.

Prussian blue A paste or liquid used to show a contact area.

Psychrophilic bacteria A group of bacteria that grow and thrive in temperatures below 68 °F (20 °C).

pH value A convenient method of expressing small differences in the acidity or alkalinity of solutions. Neutrality = pH 7; lower values indicate increasing acidity and higher values indicate increasing alkalinity.

Potable water Water fit for human consumption.

Primary waste treatment Mechanical separation of solids, grease, and scum from wastewater. With the aid of flocculating agents, primary treatment can eliminate 50–65% of the suspended solids. Solids removed by primary treatment may comprise as much as 30–40% of the original BOD of the water.

Potential to emit (PTE) Capacity (usually measured in tons per year) of a pollution source to release a particular pollutant or class of pollutants.

Purification The removal, by natural or other methods, of pollution from a given medium. The total or partial removal of unwanted constituents from a gaseous medium.

Putrefaction Biological decomposition of organic matter resulting in the production of foul-smelling products associated with anaerobic conditions.

Quality assurance/quality control (QA/QC) A system of procedures, checks, audits, and corrective actions to ensure that all research design and performance, environmental monitoring and sampling, and other technical and reporting activities achieve the program's desired data quality objectives (DQOs).

Rain water Water arising from atmospheric precipitation, which has not yet collected soluble matter from the earth.

Rated flow The gas flow rate through a separator either as stated by the manufacturer for defined conditions of use or as agreed between the interested parties for a particular installation.

Raw sewage Untreated sanitary wastewater.

Raw wastewater Plant influent or wastewater before any treatment.

Raw water Water taken from natural sources, i.e., water wells or surface water.

Reagent A substance which takes part in a chemical reaction and is used to detect and measure another substance.

Recalcination A lime-recovery process in which the calcium carbonate in sludge is converted to lime by heating to 1,800 °F (980 °C).

Recarbonation A process in which carbon dioxide is bubbled into the water being treated to lower the pH.

Receiving body A stream, lake, or other waterways into which treated or untreated waste is discharged.

Receiving water A stream, river, lake, ocean, or other surface or groundwater into which treated or untreated wastewater is discharged. It means rivers, lakes, or other water sources that receive treated or untreated wastewaters.

Reciprocating internal combustion engine (RICE) An engine in which air and fuel are introduced into cylinders, compressed by pistons, and ignited by a spark plug or by compression. Combustion in the cylinders pushes the pistons sequentially, transferring energy to the crankshaft, causing it to rotate.

Reclaimed water Reusable wastewater from wastewater treatment such as tertiary treatment of wastewater in biological and other systems.

Reducing agent Any substance, such as base metal (iron) or the sulfide ion, that will readily donate (give up) electrons. The opposite is an oxidizing agent.

Reduction Reduction is the addition of hydrogen, the removal of oxygen, or the addition of electrons to an element or compound. Under anaerobic conditions (no dissolved oxygen present), sulfur compounds are reduced to odor-producing hydrogen sulfide (H_2S) and other compounds.

Reformulated gasoline Specially refined gasoline with low levels of smog-forming volatile organic compounds (VOCs) and low levels of hazardous air pollutants.

Refractory index A measure of the ability of a substance to be biodegraded by bacterial activity.

Refractory A material having the ability to retain its shape and chemical composition when subjected to high temperatures, or the area of an incinerator or similar equipment which contains the high temperatures.

Ringelmann chart A series of charts, numbered 0 to 5, that simulate various smoke densities by presenting different percentages of black. A Ringelmann No. 1 is equivalent to 20 % black; a Ringelmann No. 5 is 100 % black. They are used for measuring the opacity or equivalent obscuration of smoke arising from stacks and other sources by matching the actual effluent with the various numbers, or densities, indicated by the charts.

Remediation Act or process of correcting a fault or deficiency without changing system structure or form.

Residual chlorine The amount of free and/or available chlorine remaining after a given contact time under specified conditions.

Residuals The solids generated and/or retained during the treatment of wastewater. They include trash, rags, grit, sediment, sludge, biosolids, septage, scum, grease, as well as those portions of treatment systems that have served their useful life and require disposal such as the sand or peat from a filter. Because of their different characteristics, management requirements can differ as stipulated by the appropriate regulations.

Residual saturation The amount of water or oil remaining in the voids of a porous medium and held in an immobile state by capillary and dead-end pores.

Respiration The process in which an organism uses oxygen for its life processes and gives off carbon dioxide.

Respirator A device designed to protect the wearer from a hazardous atmosphere.

Retardation Preferential retention of contaminant movement in the subsurface resulting from adsorption processes or solubility differences.

Retention time The time water, sludge, or solids are retained or held in a clarifier or sedimentation tank.

Return air Air entering a space from an air-conditioning, heating, or ventilating apparatus.

Return sludge The recycled sludge in a publicly owned treatment works (POTW) that is pumped from a secondary clarifier sludge hopper to the aeration tank.

Return sludge ratio (R/Q) The ratio of the return sludge flow to the wastewater flow.

Reuse The use of water or wastewater after it has been discharged and then withdrawn by another user.

Riparian of or pertaining to the banks of a body of water.

Rip rap Protective covering, often stone or coarse gravel, for earthen slopes to prevent erosion. Broken rock, cobbles, or boulders placed on earth surfaces, such as the face of a dam or the bank of a stream for the protection against erosive forces, such as flow velocity and waves.

Rising sludge Rising sludge occurs in the secondary clarifiers of activated sludge plants when the sludge settles to the bottom of the clarifier, is compacted, and then starts to rise to the surface, usually as a result of denitrification.

Risk assessment An evaluation of risk that estimates the relationship between exposure to a harmful substance and the likelihood that harm will result from that exposure.

Runoff Runoff is that part of rainfall which flows off the surface to reach a sewer or river.

Runoff reduction The process whereby practices are implemented to minimize the quantity of stormwater runoff generated and/or attenuate runoff near its source using storage, infiltration, and/or uptake by vegetation.

Runoff volume Amount of precipitation (and/or irrigation) minus surface storage, infiltration, evapotranspiration, and interception, which exits as a defined area.

Run-On Any rain water, leachate, or other liquids that drain onto any waste treatment area.

Salinity Quality of water based on its salt content; seawater contains approximately 18,000 parts per million of salt.

Saltwater Classifications: (See also Classifications) Class SA: suitable for commercial shellfishing and all other tidal saltwater uses.Class SB: saltwaters protected for primary recreation, which includes swimming on a frequent or organized basis, and all Class SC uses.Class SC: saltwaters protected for secondary recreation, fishing, and propagation and survival of aquatic life; all saltwaters are classified to protect these uses at a minimum.

Sampling line The line across the sampling plane on which the sampling points are located.

Sampling plane The cross-section of the duct in which all sampling is carried out.

Sampling point A specific location on a sampling line at which an individual sample is extracted. A specific location on a sampling line at which an individual sample is extracted.

Sampling position A place in a duct where sampling is carried out.

Sand Soil particle between 0.05 and 2.0 mm in diameter.

Sanitary collection system The pipe system for collecting and carrying liquid and liquid-carried wastes from domestic sources to a wastewater treatment plant.

Sanitary sewer A pipe or conduit (sewer) intended to carry wastewater or waterborne wastes from homes, businesses, and industries to the publicly owned treatment works (POTW). Stormwater runoff or unpolluted water should be collected and transported in a separate system of pipes or conduits (storm sewers) to natural watercourses.

Sanitary wastewater (domestic) Wastewater, including toilet, sink, shower, and kitchen flows, originating from human domestic activities.

Saturated zone The zone of the soil in which all space between the soil particles is occupied by water, including the capillary zone.

SBA Small Business Administration

SBEAP Small Business Environmental Assistance Program

SBDC Small Business Development Center

Scale A combination of mineral salts and bacterial accumulation that sticks to the inside of a collection pipe under certain conditions. Scale, in extreme growth circumstances, creates additional friction loss to the flow of water. Scale may also accumulate on surfaces other than pipes.

Screen 1. Porous material or mesh configured as a plate or cylinder that allows the passage of particles smaller than particular size (e.g., an effluent screen), according to a specific flow/pressure relationship; a screen has area but no depth with respect to flow; 2. use of a porous material or mesh in order to separate particles by size.

Scrubber An air pollution control device that uses a high energy liquid spray to remove aerosol and gaseous pollutants from an air stream. The gases are removed either by absorption or by chemical reaction.

Scum A layer or film of foreign matter (such as grease, oil) that has risen to the surface of water or wastewater; a residue deposited on the ledge of a sewer, channel, or wet well at the water surface; a mass of solid matter that floats on the surface. Usually, fatty material in wastewater that floats.

Seals (Hydraulic Seals) Seals (hydraulic seals) are used to isolate various parts of a sewer system, preventing vapor travel and spread of fire or explosion.

Secondary clarifier A wastewater treatment device which consists of a rectangular or circular tank that allows those substances not removed by previous treatment processes that settle or float to be separated from the wastewater being treated.

Secondary particulate Particles that usually form over several hours or days and attain aerodynamic diameters between 0.1 and 1 Âμm. Several of these particles, particularly those containing ammonium nitrate, are volatile and transfer mass between the gas and particle phase to maintain a chemical equilibrium.

Secondary settlement tank A thank in which settable solids or humus is separated from the effluent flowing through it from biological filters or an activated sludge plant.

Secondary standard A pollution limit based on environmental effects such as damage to property, plants, or visibility. Secondary standards are set for criteria air pollutants.

Secondary treatment A wastewater treatment process used to convert dissolved or suspended materials into a form more readily separated from the water being treated. Usually, the process follows primary treatment by sedimentation. The process commonly is a type of biological treatment process followed by secondary clarifiers that allow the solids to settle out from the water being treated.

Secondary waste treatment Processing by various types of systems that employ aeration and biological oxidation stages to decompose dissolved and colloidal organic contaminants (inorganic plant nutrients may also be partially removed).

Secondary wastewater treatment Biological removal or organics and solids from wastewater. Secondary wastewater effluent limits are generally 30 mg/L BOD5 and 30 mg/L of TSS.

Sediment Soil that has washed or eroded from a land surface. Particles of mud, sand, clay, silt, and organic matter transported and deposited by water.

Sedimentation Settling of solid material out of a liquid, typically accomplished by reducing the velocity of the liquid below the point at which it can transport the suspended material; may be enhanced by coagulation and flocculation. Separation under the action of gravity of particles from the fluid in which they are suspended. The process of settling and deposition, under the influence of gravity, of suspended matter carried by water or wastewater.

Sedimentation (wastewater) The process of settling and depositing of suspended matter carried by wastewater. Sedimentation usually occurs by gravity when the velocity of the wastewater is reduced below the point at which it can transport the suspended material.

Sedimentation basin Clarifier, Settling Tank. A tank or basin in which wastewater is held for a period of time during which the heavier solids settle to the bottom and the lighter materials float to the water surface.

Sedimentation tanks Provide a period of quiescence during which suspended waste material settles to the bottom of the tank and is scraped into a hopper and pumped out for disposal. During this period, floatable solids (fats, oils) rise to the surface of the tank and are skimmed off into scum pipes for disposal.

Seed sludge In wastewater treatment, seed, seed culture, or seed sludge refers to a mass of sludge which contains populations of microorganisms. When a seed sludge is mixed with wastewater or sludge being treated, the process of biological decomposition takes place more rapidly.

Self-purification The natural processes of purification in a polluted body of water.

Separator An apparatus for separating, from a gaseous stream in which they are suspended or mixed, solid particles (filter and dust separator), liquid particles (filter and droplet separator), or gases (gas purifier). An underground sewage disposal tank, generally installed to treat the wastewaters from an individual home, in which a continuous flow of waste material is decomposed by anaerobic (in the absence of oxygen) bacteria.

Septage Material accumulated in a pretreatment system or privy—the mat of grease and scum on the surface of septic tanks, the accumulated sludge at the bottom of tanks and the sewage present at the time of pumping.

Septic (wastewater) A condition produced by anaerobic bacteria. If severe, the wastewater produces hydrogen sulfide, turns black, gives off foul odors, contains little or no dissolved oxygen, and the wastewater has a high oxygen demand.

Septic tank A system sometimes used where wastewater collection systems and treatment plants are not available. The system is a settling tank in which settled sludge and floatable scum are in intimate contact with the wastewater flowing through the tank and the organic solids are decomposed by anaerobic bacterial action. Used to treat wastewater and produce an effluent that flows into a subsurface leaching (filtering and disposal) system where additional treatment takes place. Also referred to as an "interceptor"; however, the preferred term is "septic tank."

Septic tank effluent pump (STEP) system A facility where effluent is pumped from a septic tank into a pressurized collection system which may flow into a gravity sewer, treatment plant, or subsurface leaching system.

Septicity The condition in which organic matter decomposes to form foul-smelling products associated with the absence of free oxygen. If severe, the wastewater produces hydrogen sulfide, turns black, gives off foul odors, contains little or no dissolved oxygen, and the wastewater has a high oxygen demand.

Series operation Wastewater being treated flows through one treatment unit and then flows through another similar treatment unit.

Settling Process of subsidence and deposition of suspended matter carried by a liquid, typically accomplished by reducing the velocity of the liquid below the point at which it can transport the suspended material.

Settling time Time during which suspended, aggregated, precipitated, or colloidal substances settle by gravity.

Sewage Sewage is the fluid discharged from medical, domestic, and industrial sanitary appliances. The waterborne wastes of a community. The used water and added waste of a community which is carried away by drains and sewers. The used household water and water-carried solids that flow in sewers to a wastewater treatment plant. The preferred term is "wastewater."

Sewage effluent Treated sewage discharged from a sewage treatment works.

Sewage system Sewage system is any of several drainage systems for carrying surface water and sewage for disposal.

Sewer Sewer is an underground pipe or open channel in a sewage system for carrying water or sewage to a disposal area. A pipe or conduit that carries wastewater or drainage water. The term "collection line" is often used also.

Sewerage Sewerage is a system of sewers and ancillary works to convey sewage from its point of origin to a treatment works or other place of disposal.

Site Works or plant where sampling is to be carried out.

Sludge The accumulated suspended solids of sewage deposited in tanks or basins. A mixture of solids and water produced during the treatment of wastewater. Biosolids remaining after secondary or tertiary treatment. Sludge may be applied to agricultural fields as a soil amendment, composted or palletized. Or The settleable solids separated from liquids during processing or the deposits of foreign materials on the bottoms of streams or other water bodies.

Sludge dewatering Removing the remaining water from sludge for reuse and to lighten the sludge for reuse or disposal.

Sludge digestion The process of changing organic matter in sludge into a gas or a liquid or a more stable solid form. These changes take place as microorganisms feed on sludge in anaerobic (more common) or aerobic digesters.

Sludge gasification A process in which soluble and particulate organic matter are converted into gas by anaerobic decomposition. The resulting gas bubbles can become attached to the settled sludge and cause large clumps of sludge to rise and float on the water surface.

Sludge return Process that sends the material (sludge) settled in a clarifier back to a septic or processing tank for further treatment or to maintain adequate microbial populations for treatment.

Sludge volume index (SVI) This is a calculation used to indicate the tendency of activated sludge solids (aerated solids) in the secondary clarifier to thicken or to become concentrated during the sedimentation/thickening process. To determine SVI, allow a mixed liquor sample from the aeration basin to settle for 30 min. Also determine the suspended solid concentration for a sample of the same mixed liquor. Calculate SVI by dividing the measured (or observed) wet volume (mL/L) of the settled sludge by the dry weight concentration of MLSS in grams/L. When mixed liquor has an SVI well above 100 mL/gram of solids, it tends to form a thin slurry or billowing sludge blanket or to form bulky sludge.

Smoke A visible aerosol resulting from combustion. Note: In some literature, smoke is referred to quantitatively in terms of a Ringelmann number, a smoke shade, a darkness of stain, or a mass of deposit collected(see fume, fumes, and especially smoke, for approximate terms). The whole of the combustion gases and the particles entrained by them (by extension, also the gases charged by particles resulting from a chemical process or from a metallurgical operation).

Smog A mixture of pollutants, principally ground-level ozone, produced by chemical reactions in the air involving smog-forming chemicals. A major portion of smog-formers comes from burning petroleum-based fuels such as gasoline. Other smog-formers, volatile organic compounds, are found in products such as paints and solvents. Smog can harm health, damage the environment, and cause poor visibility. Major smog occurrences are often linked to heavy motor vehicle traffic, sunshine, high temperatures and calm winds, or temperature inversion (weather condition in which warm air is trapped close to the ground instead of rising).

Sound Sound is a form of energy emitted by a vibrating body and on reaching the ear it causes the sensation of hearing through nerves.

Source Any place or object from which pollutants are released. A source can be a power plant, factory, dry cleaning business, gas station, or a farm. Cars, trucks, and other motor vehicles are sources. Consumer products and machines used in industry can also be sources.

Softening The removal of most of the calcium and magnesium ions from water.

Soft liquid A liquid with a vapor pressure above the prevailing atmospheric pressure, e.g., ethylene, NGL, LPG.

Soil 1. Unconsolidated mineral and/or organic material on the immediate surface of the earth that serves as a medium for the growth of plants; 2. unconsolidated mineral or organic matter on the surface of the earth that has been subjected to and shows effects of pedogenic and environmental factors of climate (including water and temperature effects), and macro- and microorganisms, conditioned by relief, acting on parent material over a period of time.

Soil sorption coefficient A measure of the preference of an organic chemical to leave the dissolved aqueous phase in the soil and become attached or adsorbed to soil particles as organic carbon.

Solids, settleable Suspended solids that will settle out of suspension within a specified period of time, expressed in milliliters per liter (mL/L).

Solids, suspended That portion of total solids that is retained on a filter of 2.0 μm (or smaller) nominal pore sized under specified conditions.

Solids, total (TS) Material residue left in a vessel after evaporation of a sample subsequent to drying to a constant weight in an oven at 217–221 °F (103–105 °C), which includes total suspended solids (TSS) and total dissolved solids (TDS), typically expressed in mg/L.

Solids, total dissolved (TDS) Material that passes through a filter of 2.0 μm (or smaller) nominal pore size, evaporated to dryness in a weighed dish and subsequently dried to constant weight at 180 °C, typically expressed in mg/L.

Solids, total suspended (TSS) Measure of all suspended solids in a liquid, typically expressed in mg/L; to measure, a well-mixed sample is filtered through a standard glass-fiber filter and the residue retained on the filter is dried to a constant weight at 217–221 °F (103–105 °C); the increase in the weight of the filter represents the amount of total suspended solids.

Soil test Chemical analysis of soil to determine the need for fertilizers or amendments for species of plant being grown.

Solid waste Solid waste is any garbage, refuse, or sludge from a waste treatment plant, water supply treatment plant, or air pollution control facility, or other discarded material, including solids, liquids, semisolids, or contained gaseous material resulting from industrial, commercial, mining, and agricultural operations and from community activities. In addition, solid waste does not include radioactive source, special nuclear, and/or their by-product material.

Solubility Amount of a substance that will dissolve in a given amount of another substance, typically water.

Soot Agglomerates of carbonaceous particles formed by incomplete combustion and deposited before emission.

Sorption The removal of an ion or molecule from solution by adsorption and/or absorption; term often used when the exact nature of the mechanism of removal is not known.

Special order by consent (SOC) An administrative order entered by the Environmental Management Commission and an NPDES discharger which in some way modifies limitations of an NPDES permit by consent of both parties and provides interim limitations and conditions.

Springing Springing is separation of acid oils, either phenolic or naphthenic, by neutralization of spent caustic solutions. The acid oils are known as "sprung acids."

Stationary source A place or object from which pollutants are released which stays in place. Stationary sources include power plants, gas stations, incinerators, and houses.

Sterilization A process which inactivates or removes all living organisms (including vegetative and spore forms) as well as viruses.

Storm drain A pipeline or channel system that carries surface water and/or runoff to public waters, but does not feed into sewer system.

Storm sewage A mixture of sewage and the surface water arising from heavy rainfall or melting snow (ice).

Stormwater Stormwater is rain water discharged from a catchment area as a result of a storm.

Stormwater management facility A device that controls stormwater runoff and changes the characteristics of that runoff including, but not limited to, the quantity and quality, the period of release, or the velocity of flow.

Stormwater management plan A document describing how existing runoff characteristics will be affected by a land development project and methods for complying with the requirements of the local program.

Stormwater runoff (SRO) The pulse of surface water following a rainstorm. The water carries sediment, gas, oil, animal feces, glass, and other waste from the watershed to receiving waters creating a difficult urban/suburban wastewater problem.

Stratification The existence or formation of distinct layers in a body of water identified by thermal or salinity characteristics or by differences in oxygen or nutrient content.

Stratigraphy The study of original succession and age of subsurface layers; dealing with their form, distribution, com position, and physical and chemical properties.

Stream buffers (riparian buffers) The zones of variable width which are located along both sides of a stream and are designed to provide a protective natural area along a stream corridor.

Sublaterals Sublaterals are sewer branches [min. 150 mm (6 inch) diameter nominal size] collecting effluents from catch basins and convey it to the laterals.

Subsurface flow wetland (SF) A type of constructed wetland in which primarily treated waste flows through deep gravel or other porous substrate planted with wetland vegetation. The water is not exposed to the air, avoiding problems with odor and direct contact.

Subsurface soil absorption A process that utilizes the soil to treat and dispose of effluent from the treatment works.

Sulfur dioxide A criteria air pollutant. Sulfur dioxide is a gas produced by burning coal, most notably in power plants. Some industrial processes, such as production of paper and smelting of metals, produce sulfur dioxide. Sulfur dioxide is closely related to sulfuric acid, a strong acid. Sulfur dioxide plays an important role in the production of acid rain.

Supply water Water, which usually has been treated, that passes into a distribution network or a service reservoir.

Surface loading One of the guidelines for the design of settling tanks and clarifiers in treatment plants. Used by operators to determine whether tanks and clarifiers are hydraulically (flow) over- or underloaded. Also called overflow rate.

Surface runoff The precipitation that cannot be absorbed by the soil and flows across the surface by gravity. The water that reaches a stream by traveling over the soil surface or falls directly into the stream channels, including not only the large permanent streams but also the tiny rills and rivulets. Water that remains after infiltration, interception, and surface storage has been deducted from total precipitation.

Surface water Surface water is natural rain water from the ground surface, paved areas and roofs plus occasional courtyard and car washing wastewaters and incidental firefighting water.

Surfactant Abbreviation for surface-active agent. The active agent in detergents that possesses a high cleaning ability.

Susceptibility to separation (STS) number STS number is the oil contents in parts per million, of the suspend water after the specified settling period.

Suspended solids Solids physically suspended in sewage that can be removed by proper laboratory filtering.

Suspended solids (SS) in sewage Solid in suspension in sewage liquors as measured by filtration either through a glass-fiber filter paper followed by washing and drying at 105 °C, or by centrifuging followed by washing and removal of the supernatant liquid.

Suspended growth processes Wastewater treatment processes in which the microorganisms and bacteria treating the wastes are suspended in the wastewater being treated. The wastes flow around and through the suspended growths. The various modes of the activated sludge process make use of suspended growth reactors. These reactors can be used for BOD removal, nitrification, and denitrification.

Suspended solids 1. Solids that either float on the surface or are suspended in water, wastewater, or other liquids and which are largely removable by laboratory filtering. 2. The quantity of material removed from water in a laboratory test, as prescribed in standard methods for the Examination of Water and Wastewater and referred to as total suspended solids dried at 103° to 105°C.

Suspension A two-phase system in which one phase, known as the "dispersed medium," is distributed throughout the other, known as the " dispersion medium."

Swamp waters Waters with low velocities and other natural characteristics that differ from other surface waters.

Synthetic minor permit A permit with practically, enforceable conditions issued to a facility. These conditions limit the amount of regulated pollutant emissions so the permitted amount of actual emissions from the facility is below potential major source emission levels. These major source thresholds are usually 100 tons per year under Title V, 50 tons per year VOC/NO in the non-attainment area for ozone; 100/250 tons per year for P.D; 10 tons per year for a single hazardous air pollutant; 25 tons per year for a combination of all hazardous air pollutants; and 50 tons per year for a new source review.

Total maximum daily load (TMDL) The sum of the allowable loads of a single pollutant from all contributing point and non-point sources. It is a calculation of the maximum amount of a pollutant that a water body can receive on a daily basis and still meet water quality standards; and an allocation of that amount to the pollutant's sources.

Tank An artificial container in which liquids are held or detained.

Tank, dosing Tank or compartment which provides storage of effluent and contains a device (pump or siphon) and associated appurtenances used to convey effluent to another pretreatment process or a final treatment and dispersal component.

Tank, flow equalization Dosing tank that provides storage of effluent and uses timed dosing to allow for uniform delivery to a subsequent component over time, usually a day or more; *also known as a* surge tank.

Tank, processing Term applied to a septic tank when it is configured to receive a combination of raw sewage and recirculated effluent in order to enhance nitrogen removal.

Tank, pump Dosing tank which provides storage of effluent and houses a pump and associated appurtenances used to convey effluent to another pretreatment process or a final treatment and dispersal component.

Tank, recirculation Dosing tank that mixes effluent from two or more components within the treatment train and allows a portion of partially treated effluent to pass through one or more treatment components again.

Tank, septic Water-tight, covered receptacle for treatment of sewage; receives the discharge of sewage from a building, separates settleable and floating solids from the liquid, digests organic matter by anaerobic bacterial action, stores digested solids through a period of detention, allows clarified liquids to discharge for additional treatment and final dispersal, and attenuates flows.

Tank, siphon Dosing tank or compartment which provides storage of effluent and contains a siphon to convey effluent from the tank to another pretreatment process or to a final treatment and dispersal component.

Technical integrity The state of a system which exists when, under specified operating conditions, there is no foreseeable risk of its failure endangering people, the environment, or asset value.

Temperature controller Device which responds directly or indirectly to deviation from a desired temperature by actuating a control or initiating a control sequence.

Temperature sensor A device that opens and closes a switch in response to changes in the temperature. This device might be a metal contact, or a thermocouple that generates minute electric current proportional to the difference in heat, or a variable resistor whose value changes in response to changes in temperature.

Tertiary treatment Any process of water renovation that upgrades treated wastewater to meet specific reuse requirements. May include general cleanup of water or removal of specific parts of wastes insufficiently removed by conventional treatment processes. Typical processes include chemical treatment and pressure filtration. Also called "advanced waste treatment."

Tertiary treatments (Effluent Polishing) A furtherer stayed of treating sewage by removing suspended solids. Consequential removal of residual BOD may occur. The use of filtration to remove microscopic particles from wastewater that has already been treated to a secondary level.

Tertiary waste treatment Following secondary treatment, the clarified effluent may require additional aeration and/or other chemical treatment to destroy bacteria remaining from the secondary treating stage and to increase the content of dissolved oxygen needed for oxidation of the residual BOD. Tertiary treatment can also be used to remove nitrogen and phosphorous.

Tertiary wastewater treatment (Advanced) Biological or chemical polishing of wastewater to remove organics, solids, and nutrients. Tertiary wastewater effluent limits are generally 10 mg/L BOD5 and 10 mg/L TSS.

Test flow The gas flow rate through a separator during a rig test or a site test. This flow, which can differ from the rated flow, shall be specified or, failing this, agreed between the interested parties.

Thermocline The layer in a thermally stratified body of water in which the temperature gradient is at a maximum.

Threshold limit values (TLVs) Refer to airborne concentration of substances and represent conditions under which it is believed that nearly all workers may be repeatedly exposed day after day without adverse health effects. Refers to maximum concentration of substances which could be discharged in wastewater issued by the national standard of environmental protection agency for industrial waste.

Thickening Treatment to remove water from the sludge mass to reduce the volume that must be handled.

Tidal saltwater Tidal waters that generally have a natural chloride ion content in excess of 500 parts per million; includes all waters assigned S classifications by the Environmental Management Commission.

Toe walls Toe walls are raised curbs which control spillage and drainage of storm, process, and fire water.

Top water level (TWL) The maximum water level in a settlement tanks an aeration tank or a sludge storage tank.

Total dissolved solids Total dissolved solids (TDS) is the combined total of all dissolved solids in wastewater, both organic and inorganic and very fine, such as colloidal minerals. Generally, particles must be smaller than two micrometers to be considered a dissolved solid. For example, salt dissolved in water is a dissolved solid. Therefore, TDS will "survive" screening or other coarse filtration.

Total maximum daily load (TMDL) The total waste (pollutant) loading from point and non-point sources that a water body can assimilate while still maintaining its water quality classification and standards. The maximum amount of a pollutant that a water body can receive daily without violating water quality standards; includes best estimates of pollution from non-point sources, natural background sources, point sources, and a margin of safety; can also be defined as the strategy which is implemented to reduce or eliminate the impact of pollution.

Total organic carbon (TOC) TOC is a measure of the amount of carbon in a sample originating from organic matter only. The test is run by burning the sample and measuring the CO_2 produced.

Total solids The total amount of solids in solution and suspension.

Total suspended particulates Particles of solid and liquid matter suspended in air. TSP is collected on filtration media and analyzed by weight only. Particle sizes represented by the method are up to 100 Âμm in aerodynamic diameter.

Total suspended solids (TSS) Concentration of all substances suspended in water (solids remaining after filtering of a water sample).

Toxic pollutant Pollutants or combinations of pollutants, including disease-causing agents, which after discharge and upon exposure, ingestion, inhalation, or assimilation into any organism, either directly from the environment or indirectly by ingestion through food chains, will, on the basis of information available to the administrator of EPA, cause death, disease, behavioral abnormalities, cancer, genetic mutations, physiological malfunctions, (including malfunctions in reproduction), or physical deformations, in such organisms or their offspring. Toxic pollutants also include those pollutants listed by the administrator under CWA Section 307(a)(1) or any pollutant listed under Section 405(d) which relates to sludge management.

Toxicity test A procedure to determine the toxicity of a chemical or an effluent using living organisms. A toxicity test measures the degree of effect on exposed test organisms of a specific chemical or effluent.

Treatability How treatable a water sample is with a given substance.

Treated sewage Sewage that has received partial or complete treatment for the removal and mineralization of organic and other material.

Treatment Method, technique, or process designed to remove solids and/or pollutants from wastewater.

Treatment, aerobic Digestion of organic matter in an environment containing molecular (or dissolved) oxygen (O_2).

Treatment, advanced secondary Level of treatment that achieves 95 % reduction in BOD and TSS, generally to levels below 10 mg/L.

Treatment, anaerobic Digestion of organic matter in an environment without molecular (or dissolved) oxygen (O_2).

Treatment, biological Process involving the metabolic activities of bacteria and other microorganisms in the breakdown of complex organic materials into simpler, more stable substances.

Treatment, chemical Process involving the addition of chemicals to obtain a desired result, such as precipitation, coagulation, flocculation, pH adjustment, disinfection, or sludge conditioning.

Treatment, physical Treatment which involves only physical means of solid–liquid separation, such as filtration, flotation, and sedimentation; chemical and biological reactions do not play an important role in physical treatment.

Treatment, primary Physical treatment processes involving removal of particles, typically by settling and flotation with or without the use of coagulants (e.g., a grease interceptor or a septic tank provides primary treatment).

Treatment, secondary Biological and chemical treatment processes designed to remove organic matter; a typical standard for secondary effluent is BOD and TSS less than or equal to 20 mg/L each on a 30-day average basis.

Treatment, tertiary Advanced treatment of wastewater for enhanced organic matter removal, pathogen reduction, and nutrient removal; typical standards for tertiary effluent vary according to regulatory requirements.

Treatment works Any device or system used in the storage, treatment, disposal, or reclamation of sewage and industrial wastes, including but not limited to pumping, power and other equipment and appurtenances, septic tanks and any works, including land that are or will be an integral part of the treatment process or used for ultimate disposal of residues or effluent resulting from such treatment.

Tributary A stream or river that flows into a larger stream or river. A body of water that drains into another, usually larger, water body.

Tributary strategies A state watershed initiative, Virginia's Tributary Strategy Program, which requires the development of strategies and written plans to restore water quality and living resources of the Chesapeake Bay and its tributaries.

Trickling filter An aerobic biological wastewater treatment process used as secondary treatment of sewage. Effluent from the primary clarifier is distributed over a bed of rocks. As the liquid trickles over the rocks, a biological growth on the rocks breaks down the organic matter in the sewage. The effluent is then taken to a clarifier to remove biological matter coming from the filter.

Trout waters Freshwaters protected for natural trout propagation and survival of stocked trout.

Total suspended solids (TSS) Total suspended solids in wastewater. As the name implies, the total solid particles that are suspended (as opposed to dissolved) in the wastewater. TSS must be filtered out, flocculated, digested, and so on for the removal in the treatment of wastewater. Though not necessarily pollutants, TSS is considered to be a measure of pollutants in water by the EPA in the United States.

Turbidity The cloudy appearance of water caused by the presence of suspended and colloidal matter. In the Handbook on Wastewater Management waterworks field, a turbidity measurement is used to indicate the clarity of water. Technically, turbidity is an optical property of the water based on the amount of light reflected by suspended particles. Turbidity cannot be directly equated to suspended solids because white particles reflect more light than dark-colored particles and many small particles will reflect more light than an equivalent large particle. Any finely divided, insoluble impurities that mar the clarity of the water. A measure of the clarity of water. Typically, turbidity is measured by determining light transmission through the water.

Turbidity meter An instrument for measuring and comparing the turbidity of liquids by passing light through them and determining how much light is reflected by the particles in the liquid. The normal measuring range is 0–100 and is expressed as nephelometric turbidity units (NTUs).

Turbidity units (TU) Turbidity units are a measure of the cloudiness of water. If measured by a nephelometric (deflected light) instrumental procedure, turbidity units are expressed in nephelometric turbidity units (NTU) or simply TU. Those turbidity units obtained by visual methods are expressed in Jackson turbidity unit (JTU) which is a measure of the cloudiness of water; they are used to indicate the clarity of water. There is no real connection between NTUs and JTUs. The Jackson turbidimeter is a visual method, and the nephelometer is an instrumental method based on deflected light.

Ultrasonics The sound of frequency more than 20,000 HZ

Ultraviolet disinfection (UV) A disinfection method in which final wastewater effluent is exposed to ultraviolet light to kill pathogens and microorganisms.

Ultraviolet B (UVB) A type of sunlight. The ozone in the stratosphere filters out ultraviolet B rays and keeps them from reaching the Earth. Ultraviolet B exposure has been associated with skin cancer, eye cataracts, and damage to the environment. Thinning of the ozone layer in the stratosphere results in increased amounts of ultraviolet B reaching the Earth.

Ultra filtration A membrane filters process used for the removal of some organic compounds in an aqueous (watery) solution.

Unconfined aquifer An aquifer that is under atmospheric pressure. It is usually the uppermost aquifer in the subsurface with its upper limit being the water table.

Unit or units Unit or units refer to one or all process, offsite and/or utility units and facilities as applicable to form a complete operable refinery and/or complex.

Unit operations physical, Treatment methods in which the application of physical forces predominates as a means for the removal of wastewater constituents; includes flocculation, sedimentation, flotation, filtration, screening, mixing, and gas transfer.

Unit processes, biological Treatment methods in which the removal or conversion of constituents is brought about by biological activity; primarily used to remove the biodegradable organic constituents through conversion to cell tissue or gases; also used to remove nutrients (nitrogen and phosphorous).

Unit processes, chemical Treatment methods in which the removal or conversion of constituents is brought about through the addition of chemicals or by other chemical reactions; includes precipitation, adsorption, and disinfection.

Upgradient In the direction of increasing static head.

Upset An exceptional incident in which there is unintentional and temporary non-compliance with the permit limit because of factors beyond the reasonable control of the permittee. An upset does not include non-compliance to the extent caused by operational error, improperly designed treatment facilities, inadequate treatment facilities, lack of preventive maintenance, or careless or improper operation.

Urban runoff Stormwater from city streets and adjacent domestic or commercial properties that carries non-point source pollutants of various kinds into the sewer systems and receiving waters.

Vadose zone The portion of a porous medium above the water table within which the capillary pressure is less than atmospheric and the moisture content is usually less than saturation. The vadose zone includes the capillary fringe.

Vapor The gaseous phase of liquids or solids at atmospheric temperature and pressure.

Vapor pressure The equilibrium pressure exerted on the atmosphere by a liquid or solid at a given temperature. Also a measure of a substance's propensity to evaporate or give off flammable vapors. The higher the vapor pressure, the more volatile the substance.

Vapor recovery nozzles Special gas pump nozzles that reduce the release of gasoline vapor into the air gas are pumped into car tanks. There are several types of vapor recovery nozzles. Therefore, nozzles may not look the same at all gas stations. The 1990 Clean Air Act requires the installation of vapor recovery nozzles at gas stations in smoggy areas.

Vapor recovery systems Mechanical systems that collect and recover chemical vapors resulting from transfer of gasoline from operations such as tank-to-truck systems at refineries, tanker-to-pipeline systems at offshore oil operations, and pump-to-vehicle systems at gasoline stations.

Variance Permission granted for a limited time (under stated conditions) for a person or company to operate outside the limits prescribed in a regulation.

Viscosity In flowing liquids, the existence of internal friction or the internal resistance to relative motion of the fluid particles with respect to each other must be considered; this resistance is called viscosity.

Visibility A measurement of the ability to see and identify objects at different distances. Visibility reduction from air pollution is often due to the presence of sulfur and nitrogen oxides, as well as particulate matter.

Volatile oxygen demand (VOD) Compounds which under favorable conditions may participate in photochemical reaction to form oxidants, typically excludes methane and ethane.

Volatile A volatile substance is one that is capable of being evaporated or changed to a vapor at relatively low temperatures. Volatile substances also can be partially removed by air stripping. In terms of solid analysis, volatile refers to materials lost (including most organic matter) upon ignition in a muffle furnace for 60 min at 550 °C. Natural volatile materials are chemical substances usually of animals or plant origin. Manufactured or synthetic volatile materials such as ether, acetone, and carbon tetrachloride are highly volatile and not of plant or animal origin

Volatile acids Fatty acids produced during digestion that are soluble in water and that can be steam-distilled at atmospheric pressure. Also called "organic acids." Volatile acids are commonly reported as equivalent to acetic acid.

Volatile organic compounds (VOCs) Organic chemicals all contain the element carbon (C). Organic chemicals are the basic chemicals found in living things and in products derived from living things, such as coal, petroleum, and refined petroleum products. Many of the organic chemicals we use do not occur in nature, but were synthesized by chemists in laboratories. Volatile chemicals readily produce vapors at room temperature and normal atmospheric pressure. Vapors escape easily from volatile liquid chemicals. Volatile organic chemicals include gasoline, industrial chemicals such as benzene, solvents such as toluene and xylene, and tetrachloroethylene (perchloroethylene, the principal dry cleaning solvent). Many volatile organic chemicals, such as benzene, are also hazardous air pollutants.

Volatile solids Those solids in water, wastewater, or other liquids that are lost on ignition of the dry solids at 550 °C for 60 min.

Volatilization The process of transfer of a chemical from the water or liquid phase to the air phase. Solubility, molecular weight, and vapor pressure of the liquid and the nature of the air–liquid/ water interface affect the rate of volatilization.

Volume, dose 1. Amount of effluent delivered to the distribution system during a dosing event including the drainback volume, pipe fill volume, and the delivered dose volume; 2. Amount of effluent delivered as determined by the pump-on and pump-off levels in a demand dosed system.

Vulnerability assessment (water) An evaluation of drinking water source quality and its vulnerability to contamination by pathogens and toxic chemicals.

Washer Any dust separator, droplet separator, or gas purifier that depends for its operation on a liquid acting as a collecting medium.

Waste activated sludge (WAS) mg/L, The excess growth of microorganisms which must be removed from the process to keep the biological system in balance. That portion of sludge from the secondary clarifier in the activated sludge process that is wasted to avoid a buildup of solids in the system.

Wastewater A community's used water and water-carried solids (including used water from industrial processes) that flow to a treatment plant. Stormwater, surface water, and groundwater infiltration also may be included in the wastewater that enters a wastewater treatment plant. The term "sewage" usually refers to household wastes, but this word is being replaced by the term "wastewater." The liquid-borne waste products of domestic, industrial, agricultural and manufacturing activities. In a community, an average of 50–100 gallons of wastewater is generated per person per day.

Wastewater facilities The pipes, conduits, structures, equipment, and processes required to collect, convey, and treat domestic and industrial wastes, and dispose of the effluent and sludge.

Wastewater ordinance The basic document granting authority to administer a pretreatment inspection program. This ordinance must contain certain basic elements to provide a legal framework for effective enforcement.

Wastewater stabilization pond Constructed basin lined with either soil with very low permeability or a synthetic material, surrounded with berms and which contains at least three feet of wastewater which utilizes sunlight, wind or mechanical aeration, and natural bacteria to break down waste via physical, chemical, and biological processes to stabilize wastewater, typically consists of two or more basins with operational controls allowing or facilitating flow through the basins.

Wastewater treatment plant An arrangement of pipes, equipment, devices, tanks, and structures for treating wastewater and industrial wastes. A water pollution control plant.

Wastewater, industrial Water or liquid-carried waste from an industrial process resulting from industry, manufacture, trade, automotive repair, vehicle wash, business or medical, activity; this wastewater may contain toxic or hazardous constituents.

Wastewater, residential strength Effluent from a septic tank or other treatment device with a BOD5 less than or equal to 170 mg/L; TSS less than or equal to 60 mg/L; and fats, oils, and grease less than or equal to 25 mg/L.

Wastewater, raw Any wastewater leaving a source.

Wastewater reclamation Treatment or processing of wastewater to produce water of a quality appropriate for another use, including recycling or reuse; *see also*.

Wastewater recycling Reclamation process of collection and treatment of wastewater on-site for return and use back into the same site; for example, collection and reclamation of graywater from an establishment for subsequent toilet flushing in that same establishment.

Wastewater reuse Reclamation process of collection and treatment of wastewater for the deliberate application of that treated wastewater for a beneficial purpose such as turf irrigation.

Wastewater treatment system Assembly of components for collection, treatment, and dispersal of sewage or effluent.

Wastewater treatment system, cluster Wastewater treatment systems designed to serve two or more sewage-generating dwellings or facilities with multiple owners, typically includes a comprehensive, sequential land use planning component, and private ownership.

Wastewater treatment system, collector Wastewater treatment system that conveys sewage or effluent from multiple sources to a location where treatment and dispersal occur.

Wastewater treatment system, community Publicly owned wastewater treatment system for collection, treatment, and dispersal of wastewater from two or more lots, or two or more equivalent dwelling units.

Wastewater treatment system, decentralized wastewater treatment system for collection, treatment, and dispersal/reuse of wastewater from individual homes, clusters of homes, isolated communities, industries, or institutional facilities, at or near the point of waste generation.

Wastewater treatment system, individual Wastewater treatment system designed to serve one sewage-generating dwelling or facility.

Wastewater treatment system, onsite (OWTS) Wastewater treatment system relying on natural processes and/or mechanical components to collect and treat sewage from one or more dwellings, buildings, or structures and disperse the resulting effluent on property owned by the individual or entity.

Water pollution A general term signifying the introduction into water of microorganisms, chemicals, wastes, or sewage which renders the water unfit for its intended use.

Water quality criteria levels of water quality expected to render a body of water suitable for its designated use. Criteria are based on specific levels of pollutants that would make the water harmful if used for drinking, swimming, fish production, or industrial uses.

Water quality standard (WQS) A law or regulation that consists of the beneficial use or uses of a water body, the numeric and narrative water quality criteria that are necessary to protect the use or uses of that particular waterbody and an antidegradation statement.

Water quality-based effluent limit (WQBEL) A value determined by selecting the most stringent of the effluent limits calculated using all applicable water quality criteria (e.g., aquatic life, human health, and wildlife) for a specific point source to a specific receiving water for a given pollutant.

Water softening Reduction in the number of and/or removal of polyvalent cations which are the principal cause of hardness in water.

Water table Upper surface of groundwater or that level in the ground where the water is at atmospheric pressure. The uppermost surface of groundwater saturation—the level in the saturated zone at which the pressure is equal to atmospheric pressure.

Water treatment discharge By-product from a water treatment device, such as regeneration water from an ion-exchange unit, reject water from a reverse-osmosis unit, or the backwash from an iron filter.

Water well or well Any artificial opening or artificially altered natural opening, however made, by which groundwater is sought or through which groundwater flows under natural pressure or is intended to be artificially drawn; provided this definition shall not include wells drilled for exploration or production of gas or oil; building foundation investigation and construction; elevator shafts; grounding of electrical apparatus; or the modification or development of springs.

Watershed Watershed a geographic area in which water, sediment, and dissolved materials drain to a common outlet such as a point on a larger stream, a lake, an underlying aquifer, an estuary, or an ocean. A drainage area or basin in which all land and water areas drain or flow toward a central collector such as a stream, river, or lake at a lower elevation.

Weathering The process where a complex compound is reduced to its simpler component parts, transported through physical processes, or biodegraded over time.

Weir Device designed to measure or control flow; consists of a wall or obstruction of known geometric shape placed perpendicular to the direction of flow. A wall or plate placed in an open channel to regulate or measure the flow of water.

Well Hole bored or drilled into the ground.

Well, monitoring Well constructed for the purpose of determining groundwater level or constituents.

Well, water Well constructed for the purpose of extracting potable water.

Well-sorted Material of uniform size with maximum void space.

Wetlands Areas inundated or saturated by surface or groundwater at a frequency and duration to support and that, under normal circumstances, do support, a prevalence of vegetation typically adapted for life in saturated soil conditions. Coastal wetlands extend back from estuaries and include salt marshes, tidal basins, marshes, and mangrove swamps. Inland freshwater wetlands consist of swamps, marshes, and bogs.

Whole effluent toxicity (WET) The total toxic effect of an effluent measured directly with a toxicity test.

Zero air Pure air, used for calibrating air monitoring instruments. The EPA requires zero air to have less than 0.1 ppm of hydrocarbons.

Zone Portion of a component that is separately managed as a single unit.

Zone of dispersal Layers of soil or rock material surrounding the zone of treatment through which the effluent moves away from the final treatment and dispersal component.

Zone of saturation Layer in the ground in which interstitial voids (cracks, crevices, holes, etc.) are filled with water; the level at the top of this zone is the water table.

Zone of treatment Soil or fill material which removes pollutants from pretreated effluent by processes which include physical filtration of bacteria and other constituents, adsorption of viruses and bacteria by clay and organic matter, biological destruction of pathogens by soil microorganisms, sorption or precipitation of phosphorus, biochemical transformations of nitrogen compounds, and biological assimilation of phosphorus and nitrogen.

References

1. Theodore L, Bounicore AJ (1982) Air pollution control equipment: selection, design, operation, and maintenance. Prentice Hall, Englewood Cliffs
2. EPA (1974) The World air quality management standards volume II", US. Environmental Protection Agency, Office of Research and Development, EPA-650/9-75-001b
3. ISO 6584 (1981) Cleaning equipment for air and other gases—classification of dust separators
4. Taback (1996) Estimating VOC emissions from petroleum industry sources, Hal Taback Company, Sept 1996 (Course notes presented at AWMA Conference on Emission Estimation, New Orleans, LA, Sept 1996)
5. U.S. Environmental Protection Agency (EPA) (1995) AP 42 Compilation of air pollutant emission factors. United States
6. USEPA (1990) Air emissions species manual, vol 1. Volatile organic species profiles, 2nd edn. Office of Air Quality Planning and Standards, Research Triangle Park, (EPA-450/2-90-001a)
7. USEPA (1993) VOC/PM Speciation DBMS. Office of Air Quality Planning and Standards, US EPA, Research Triangle Park
8. USEPA (1995a) Protocol for equipment leak emission estimates, Office of Air Quality Planning and Standards, US EPA, Research Triangle Park, (EPA-453/R-95-017)
9. USEPA (1997), Compilation of Air Pollutant Emission Factors, vol I. Stationary sources, 5th edn. Office of Air Planning and Standards, Office of Air Quality Planning and Standards, US EPA, Research Triangle Park, 1995 (AirChief CD-ROM, 1997, Version 5.0)
10. USEPA (1997) Factor information retrieval system (FIRE), Office of Air Quality Planning and Standards, US EPA, Research Triangle Park, (AirChief CD-ROM, 1997,Version 5.0)
11. USEPA (1998) Fuel oil combustion, March 1998 Revision to AP-42 Section 1.3
12. USEPA (1998) Natural gas combustion, September 1998 Revision to AP-42 Section 1.4
13. Braddock JD (1995) Developments in regulation of air pollution from oil and gas exploration and production, SPE/EPA Exploration and Production Environmental Conference, pp 535–549
14. Schifter I, González-Macías C, Miranda A, López-Salinas E (2005) Air emissions assessment from offshore oil activities in Sonda de Campeche. Mexico Environ Monit Assess 109(1-3):135–145
15. Karbassi AR, Abbasspour M, Sekhavatjou MS, Ziviyar F, Saeedi M (2008) Potential for reducing air pollution from oil refineries. Environ Monit Assess 145(1–3):159–166
16. Baltrenas P, Baltrenaite E, Šerevičiene V, Pereira P (2011) Atmospheric BTEX concentrations in the vicinity of the crude oil refinery of the Baltic region. Environ Monit Assess 182(1-4):115–127
17. Zadakbbar O, Vatani A, Karimpour K (2008) Flare gas recovery in oil and refineries. Oil Gas Sci Technol 63(6):705–711

A. Bahadori, *Pollution Control in Oil, Gas and Chemical Plants*,
DOI: 10.1007/978-3-319-01234-6,

18. Abdulkareem AS, Odigure JO, Abenege S (2009) Predictive model for pollutant dispersion from gas flaring: a case study of oil producing area of Nigeria. Energ Sour Part A: Recovery Utilization Environ Eff 31(12):1004–1015
19. Ross JL, Ferek RJ, Hobbs PV (1996) Particle and gas emissions from an in situ burn of crude oil on the ocean. J Air Waste Manage Assoc 46(3):251–259
20. Driussi C, Jansz J (2006) Technological options for waste minimisation in the mining industry. J Cleaner Prod 14:682–688
21. Mao I-F, Chen M-R, Wang L, Chen M-L, Lai S-C, Tsai C-J (2012) Method development for determining the malodor source and pollution in industrial park. Sci Total Environ 437:270–275
22. Johnson MR, Coderre AR (2011) An analysis of flaring and venting activity in the Alberta upstream oil and gas industry. J Air Waste Manage Assoc 61(2):190–200
23. Osuji LC, Avwiri GO (2005) Flared gases and other pollutants associated with air quality in industrial areas of Nigeria: an overview. Chem Biodivers 2(10):1277–1289
24. Jou CJG, Hsieh SC (2008) Reduction of greenhouse gas emission through applying hydrogen-rich fuel on industrial boiler. Pract Periodical Hazard Toxic Radioactive Waste Manage 12(4):270–274
25. AL-Hamad K Kh, Nassehi V, Khan AR (2008) Impact of green house gases (GHG) emissions from oil production facilities at Northern Kuwait oilfields: simulated results. Am J Environ Sci 4(5):491–501
26. Coupard M, Hournac R (1985) New processing trends for reducing oil refinery pollution. Ind Environ 8(2):26–30
27. Akeredolu FA, Sonibare JA (2004) A review of the usefulness of gas flares in air pollution control. Manage Environ Qual 15(6):574–583
28. Gaffney JS, Marley NA (2009) The impacts of combustion emissions on air quality and climate—from coal to biofuels and beyond. Atmos Environ 43(1):23–36
29. Anikeev DR, Yusupov IA, Luganskii NA, Zalesov SV, Lopatin KI (2006) Effect of emissions from petroleum gas flares on the reproductive state of pine stands in the northern taiga subzone. Russ J Ecol 37(2):109–113
30. Saeki Y, Emura T (2002) Technical progresses for PVC production. Prog Polym Sci 27:2055–2131
31. Yang X, Sun M (2013) Environmental pollution and protection of oil and gas production and utilization in China. Appl Mech Mater 261–262:648–653
32. Littlejohn D, Lucas D (2003) Tank atmosphere perturbation: a procedure for assessing flashing losses from oil storage tanks. J Air Waste Manage Assoc 53(3):360–365
33. Sonibare JA, Ede PN (2009) Potential impacts of integrated oil and gas plant on ambient air quality. Energ Environ 20(3):331–344
34. Aycaguer A-C, Lev-On M, Winer AM (2001) Reducing carbon dioxide emissions with enhanced oil recovery projects: a life cycle assessment approach. Energ Fuels 15(2):303–308
35. Countess RJ, Browne D (1993) Fugitive hydrocarbon emissions from pacific offshore oil platforms: models, emission factors, and platform emissions. J Air Waste Manage Assoc 43(11):1455–1460
36. Bergerson JA, Kofoworola O, Charpentier AD, Sleep S, MacLean HL (2012) Life cycle greenhouse gas emissions of current oil sands technologies: surface mining and in situ applications. Environ Sci Technol 46(14):7865–7874
37. Snow N (2007) BP refinery leads US in carcinogenic emissions, group says. Oil Gas J 105(7):32–34
38. Viswanath RS (1994) Characteristics of oil field emissions in the vicinity of Tulsa, Oklahoma. J Air Waste Manage Assoc 44(8):989–994
39. Dadashzadeh M, Khan F, Hawboldt K, Abbassi R (2011) Emission factor estimation for oil and gas facilities. Process Saf Environ Prot 89(5):295–299

40. Khan FI, Husain T, Abbasi SA (2002) Design and evaluation of safety measures using a newly proposedmethodology "SCAP". J Loss Prev Process Ind 15:129–146
41. Tang J, Bao Z, Xiang W, Gou Q (2008) Geological emission of methane from the Yakela condensed oil/gas field in Talimu Basin, Xinjiang, China. J Environ Sci 20(9):1055–1062
42. Ba-Shammakh MS (2010) Generalized mathematical model for SO2 reduction in an oil refinery based on arabian light crude oil. Energ Fuels 24(6):3526–3533
43. Tarver GA, Dasgupta PK (1997) Oil field hydrogen sulfide in Texas: emission estimates fate. Environ Sci Technol 31(12):3669–3676
44. Karras G (2010) Combustion emissions from refining lower quality oil: what is the global warming potential? Environ Sci Technol 44(24):9584–9589
45. Croft TA (1973) Burning waste gas in oil fields. Nature 245(5425):375–376
46. Abdul-Wahab S, Ali S, Sardar S, Irfan N (2012) Impacts on ambient air quality due to flaring activities in one of Oman's oilfields. Arch Environ Occup Health 67(1):3–14
47. Gilman JB, Lerner BM, Kuster WC, De Gouw JA (2013) Source signature of volatile organic compounds from oil and natural gas operations in northeastern Colorado. Environ Sci Technol 47(3):1297–1305
48. Stratton RW, Wong HM, Hileman JI (2011) Quantifying variability in life cycle greenhouse gas inventories of alternative middle distillate transportation fuels. Environ Sci Technol 45(10):4637–4644
49. Schwartz R, Keller M (1977) Environmental factors vs. flare application. Chem Eng Prog 73(9):41–44
50. Cheng J, Luo Y (2013) Modified explosive diagram for determining gas-mixture explosibility. J Loss Prev Process Ind (in press) http://dx.doi.org/10.1016/j.jlp.2013.02.007
51. Yeh S, Jordaan SM, Brandt AR, Turetsky MR, Spatari S, Keith DW (2010) Land use greenhouse gas emissions from conventional oil production and oil sands. Environ Sci Technol 44(22):8766–8772
52. Whitcombe JM, Cropp RA, Braddock RD, Agranovski IE (2003) Application of sensitivity analysis to oil refinery emissions. Reliab Eng Syst Saf 79(2):219–224
53. Mahmoud A, Shuhaimi M, Abdel Samed M (2009) A combined process integration and fuel switching strategy for emissions reduction in chemical process plants. Energy 34(2):190–195
54. Johnson MR, Kostiuk LW, Spangelo JL (2001) A characterization of solution gas flaring in alberta. J Air Waste Manage Assoc 51(8):1167–1177
55. Skea J (1993) Market-based instruments for greenhouse gas control in the European community. Energ Convers Manage 34(9-11):789–796
56. McNeal BL, Layfield DA, Norvell WA, Rhoades JD (1969) Factors influencing hydraulic conductivity of soils in the presence of mixed-salt solutions. Soil Sci Soc Am Proc 32:187–190
57. Tchobanoglous G, Burton FL, Stensel HD (2003) Wastewater engineering: treatment and reuse, 4th edn. Metacalf & Eddy, Inc. McGraw-Hill, NY
58. Khatoonabadai A, Dehcheshmeh ARM (2006) Oil pollution in the Caspian Sea coastal waters. Int J Environ Pollut 26(4):347–363
59. Wan YS, Bobra M, Bobra AM, Maijanen A, Suntio L, Mackay D (1990) The water solubility of crude oils and petroleum products. Oil Chem Pollut 7(1):57–84
60. Özgen Karacan C, Ruiz Felicia A, Cotè Michael, Phipps Sally (2011) Coal mine methane: a review of capture and utilization practices with benefits to mining safety and to greenhouse gas reduction. Int J Coal Geol 86:121–156
61. Guo J, Li L, Liu J, Jia J (2006) Rapid detecting of chemical oxygen demand of petroleum polluted water. Speciality Petrochemicals 23(5):14–17
62. Bidleman TF, Castleberry AA, Foreman WT, Zaranski MT, Wall DW (1990) Petroleum hydrocarbons in the surface water of two estuaries in the Southeastern united states. Estuarine Coast Shelf Sci 30(1):91–109

63. Literathy P (2006) Monitoring and assessment of oil pollution in the Danube River during the transnational Joint Danube Survey. Water Sci Technol 53(10):121–129
64. Huang SZ, Zhao XZ (2013) A design of wireless sensor system for water quality monitoring of oil field. Appl Mech Mater 281:51–54
65. Okandan E, Gümrah F, Demiral B (2001) Pollution of an aquifer by produced oil field water. Energ Sour 23(4):327–336
66. Fakhru'l-Razi A, Pendashteh A, Abdullah LC, Biak DRA, Madaeni SS, Abidin ZZ (2009) Review of technologies for oil and gas produced water treatment. J Hazard Mater 170(2-3):530–551
67. Emole CE (1998) Regulation of oil and gas pollution. Environ Policy Law 28(2):103–112
68. Li KY, Kane AJ, Wang JJ, Cawley WA (1993) Measurement of biodegradation rate constants of a water extract from petroleum-contaminated soil. Waste Manage 13(3):245–251
69. Vasenko OG (1998) Environmental situation in the Lower Dnipro River Basin. Water Qual Res J Can 33(4):457–487
70. Lin D (1999) Water treatment complicates heavy oil production. Oil Gas J 97(38):76–78
71. Literathy P, Haider S, Samhan O, Morel G (1989) Experimental studie on biological and chemical oxidation of dispersed oil in seawater. Water Sci Technol 21(8-9):845–856
72. Mandke JS (1990) Corrosion causes most pipeline failures in Gulf of Mexico. Oil Gas J 88(44):40–44
73. Nadim F, Liu S, Hoag GE, Chen J, Carley RJ, Zack P (2002) A comparison of spectrophotometric and gas chromatographic measurements of heavy petroleum products in soil samples. Water Air Soil Pollut 134(1-4):97–109
74. Van De Weghe H, Vanermen G, Gemoets J, Lookman R, Bertels D (2006) Application of comprehensive two-dimensional gas chromatography for the assessment of oil contaminated soils. J Chromatogr A 1137(1):91–100
75. Pérez Pavón JL, García Pinto C, Guerrero Peña A, Moreno Cordero B (2008) Headspace mass spectrometry methodology: application to oil spill identification in soils. Anal Bioanal Chem 391(2):599–607
76. Wan C, Yang X, Du M, Xing D, Yu C, Yang Q (2010) Desorption of oil in naturally polluted soil promoted by ß-cyclodextrin. Fresen Environ Bull 19(7):1231–1237
77. Carls EG, Fenn DB, Chaffey SA (1995) Soil contamination by oil and gas drilling and production operations in Padre Island National Seashore, Texas, U.S.A. J Environ Manage 45(3):273–286
78. Lai CC, Huang YC, Wei YH, Chang JS (2009) 1Biosurfactant-enhanced removal of total petroleum hydrocarbons from contaminated soil. J Hazard Mater 167(1-3):609–614
79. Abdol Hamid HR, Kassim WMS, Hishir A, El-Jawashi SAS (2008) Risk assessment and remediation suggestion of impacted soil by produced water associated with oil production. Environ Monit Assess 145(1-3):95–102
80. Renneberg A, Dudas MJ (2002) Calcium hypochlorite removal of mercury and petroleum hydrocarbons from co-contaminated soils. Waste Manage Res 20(5):468–475
81. Sato S, Matsumura A, Urushigawa Y, Metwally M, Al-Muzaini S (1998) Type analysis and mutagenicity of petroleum oil extracted from sediment and soil samples in Kuwait. Environ Int 24(1–2):67–76
82. Jovančićević B, Antić M, Pavlović I, Vrvić M, Beškoski V, Kronimus A, Schwarzbauer J (2008) Transformation of petroleum saturated hydrocarbons during soil bioremediation experiments water. Air Soil Pollut 190(1–4):299–307
83. Ünlü K, Demirekler E (2000) Modeling water quality impacts of petroleum contaminated soils in a reservoir catchment. Water Air Soil Pollut 120(1-2):169–193
84. Liu R, Jadeja RN, Zhou Q, Liu Z (2012) Treatment and remediation of petroleum-contaminated soils using selective ornamental plants. Environ Eng Sci 29(6):494–501
85. Creighton K, Richards R (1997) Field screening technique for total petroleum hydrocarbons in soils. Pract Periodical Hazard Toxic Radioactive Waste Manage 1(2):78–83

86. Zhang W, Li J, Huang G, Song W, Huang Y (2011) An experimental study on the biosurfactant-assisted remediation of crude oil and salt contaminated soils. J Environ Sci Health Part A Toxic/Hazard Subst Environ Eng 46(3):306–313
87. Brusturean G-A, Todinca T, Perju D, Carré J, Bourgois J (2007) Soil clean up by venting: comparison between modelling and experimental VOC removal results. Environ Technol 28(10):1153–1162
88. Stokman SK, Sogorka BJ (1997) Soil contamination: dealing with petroleum spills. Chem Eng (New York) 104(1):113–116
89. Chaîneau CH, Yepremian C, Vidalie JF, Ducreux J, Ballerini D (2003) Bioremediation of a crude oil-polluted soil: biodegradation, leaching and toxicity assessments water. Air Soil Pollut 144(1-4):419–440
90. Nadim F, Hoag GE, Liu S, Carley RJ, Zack P (2000) Detection and remediation of soil and aquifer systems contaminated with petroleum products: an overview. J Petrol Sci Eng 26:169–178
91. Embar K, Forgacs C, Sivan A (2006) The role of indigenous bacterial and fungal soil populations in the biodegradation of crude oil in a desert soil. Biodegradation 17(4):369–377
92. Bahadori A, Al-Haddabi M, Vuthaluru HB (2012) Simple predictive tool estimates sodium adsorption ratio for evaluation of potential infiltration problems using reclaimed wastewater. Commun Soil Sci Plant Anal 43(19):2492–2503
93. Mathur JSB (1981) Noise control: methods of reduction, industrial effluent treatment, vol 12. Applied Science Publishers Ltd., London
94. U.S. Environmental Protection Agency (EPA) (1995) Protocol for equipment leak emission estimates. (EPA-453/R-95-017). United States
95. U.S. Environmental Protection Agency (EPA) (2007) Leak detection and repair compliance assistance guidance best practices guide. United States
96. U.S. Environmental Protection Agency (EPA) (1994) Alternative control techniques document: NOx emissions from process heaters. United States
97. Martino G (1998) Réformage catalytique. In: Leprince P (ed) Le raffinage du pétrole, tome 3, Procédés de transformation, Technip, pp 105–173
98. Travers C (1998) Isomérisation des paraffines légères. In: Leprince P (ed) Le raffinage du pétrole, tome 3,Procédés de transformation, Technip, pp 237–264
99. Henrich G, Kasztelan S (1998) Hydrotraitements. In: Leprince P (ed) Le raffinage du pétrole, tome 3,Procédés de transformation, Technip, pp 549–590
100. Decoopman F (1998) Traitement des eaux. In: Leprince P (ed) Le raffinage du pétrole, tome 3, Procédés detransformation, Technip, pp. 657–684
101. U.S. Environmental Protection Agency (EPA) (1995) Profile of Petroleum Refining Industry.Office of Enforcement and Compliance Assurance, United States
102. World Bank Group (2007) Environmental, health, and safety guidelines for petroleum refining
103. USEPA (1989) Estimating air toxic emissions from coal and oil combustion sources, Office of Air Quality Planning and Standards, US EPA, Research Triangle Park, (EPA-450/2-89-001)
104. Franken AP (ed) (1974) Community noise pollution, industrial pollution, Van Noistrand Reinhold Company, New York
105. Rao PR (1995) Noise pollution and control, encyclopedia of environmental pollution and control, vol 2. Environmedia Publications, India
106. Mathur JSB (1981) Noise control: legislation, planning and design, industrial effluent treatment, vol 2. Applied Science Publishers Ltd., London
107. Delucchi MA (2010) Impacts of biofuels on climate change, water use, and land use. Ann New York Acad Sci 1195:28–45
108. Lin T-C, Pan P-T, Cheng S–S (2010) Ex situ bioremediation of oil-contaminated soil. J Hazard Mater 176(1-3):27–34

109. Gupta MK, Srivastava RK (2010) Evaluation of engineering properties of oil-contaminated soils. J Inst Eng (India): Civ Eng Div 90:37–42
110. Kang S-W, Kim Y-B, Shin J-D, Kim E-K (2010) Enhanced iodegradation of hydrocarbons in soil by microbial biosurfactant, sophorolipid. Appl Biochem Biotechnol 160(3):780–790
111. Mao D, Lookman R, Weghe HVD, Weltens R, Vanermen G, Brucker ND, Diels L (2009) Estimation of ecotoxicity of petroleum hydrocarbon mixtures in soil based on HPLC-GCXGC analysis. Chemosphere 77(11):1508–1513
112. Duru, UI, Ossai IA, Ossai CI, Arubi IMT (2009) The after effect of crude oil spillage on some associated heavy metals in the soil, society of Petroleum Engineers. In: International petroleum technology conference, IPTC 2009, vol 2, pp 1161–1165
113. Baawain M, Al-Zidi B (2009) Petrochemicals. Water Environ Res 81(10):1664–1686
114. Mackie A, Woszczynski M, Farmer H, Walsh ME, Gagnon GA (2009) Water reclamation and reuse. Water Environ Res 81(10):1406–1418
115. Redel-Macías MD, Pinzi S, Leiva D, Cubero-Atienza AJ, Dorado MP (2012) Air and noise pollution of a diesel engine fueled with olive pomace oil methyl ester and petrodiesel blends. Fuel 95:615–621
116. Aisien FA, Chiadikobi JC, Aisien ET (2009) Toxicity assessment of some crude oil comtaminated soils in the niger delta. Adv Mater Res 62–64:451–455
117. Nistov A, Klovning R, Lemstad F, Risberg J, Ognedal TA, Haver PA, Skogesal AJ (2012) Noise reduction interventions in the Norwegian petroleum industry, society of petroleum engineers—SPE/APPEA. International conference on health, safety and environment in oil and gas exploration and production 2012: protecting people and the environment—evolving challenges, vol. 2, pp 1278–1285
118. Shevnin V, Delgado-Rodríguez O, Mousatov A, Nakamura-Labastida E, Mejía-Aguilar A (2003) Oil pollution detection using resistivity sounding. Geofisica Internacional 42(4):613–622
119. Hadad A, Cahyono D, Ardhana Putra IB, Susanto J, Djoko Rianto BU, Tjahjono EW (2009) Hearing conservation program in oil and gas company society of petroleum engineers—SPE/IATMI Asia Pacific health safety, security and environment conference and exhibition, APHSSEC 09, pp 235–240
120. Sultana S, Zhi C (2007) A review of noise impacts from offshore oil-gas production activities on the marine biota. Can Acoust Acoustique Canadienne 35(3):174–176
121. Roberts C (2008) A guideline for the assessment of low-frequency noise. Acoust Bull 33(5):31–32 + 34–36
122. Zadakbbar O, Vatani A, Karimpour K (2008) Flare gas recovery in oil and refineries. Oil Gas Sci Technol 63(6):705–711
123. Aliev TA, Guluyev GA, Pashayev FH, Sadygov AB (2012) Noise monitoring technology for objects in transition to the emergency state. Mech Syst Signal Process 27(1):755–762

Author Biography

Alireza Bahadori, Ph.D. is a research staff member in the School of Environment, Science and Engineering at Southern Cross University in Australia. He received his Ph.D. from Curtin University, Western Australia. For the better part of 20 years, Dr. Bahadori had held various process engineering positions and involved in many large-scale projects at NIOC, Petroleum Development Oman (PDO), and Clough AMEC PTY LTD.

He is the author of over 200 articles and 10 books. His books have been accepted/published by prestigious publishers such as *John Wiley & Sons, Springer, Taylor & Francis, and Elsevier*. Dr. Bahadori is the recipient of highly competitive and prestigious Australian Government's Endeavour International Postgraduate Research award as part of his research in oil and gas area. He also received top-up award from State Government of Western Australia through Western Australia Energy Research Alliance (WA:ERA) in 2009.

A. Bahadori, *Pollution Control in Oil, Gas and Chemical Plants*,
DOI: 10.1007/978-3-319-01234-6,

Index

A. Bahadori, *Pollution Control in Oil, Gas and Chemical Plants*,
DOI: 10.1007/978-3-319-01234-6,

MIX
Papier aus verantwortungsvollen Quellen
Paper from responsible sources
FSC® C105338
FSC
www.fsc.org

If you have any concerns about our products,
you can contact us on
ProductSafety@springernature.com

In case Publisher is established outside the EU,
the EU authorized representative is:
Springer Nature Customer Service Center GmbH
Europaplatz 3, 69115 Heidelberg, Germany

Printed by Libri Plureos GmbH
in Hamburg, Germany